Cambridge
International AS & A Level

Mathematics
Probability and
Statistics 1

Sophie Goldie
Series editor: Roger Porkess

HODDER
EDUCATION
AN HACHETTE UK COMPANY

Questions from the Cambridge International AS & A Level Mathematics papers are reproduced by permission of Cambridge Assessment International Education. Unless otherwise acknowledged, the questions, example answers, and comments that appear in this book were written by the authors. Cambridge Assessment International Education bears no responsibility for the example answers to questions taken from its past question papers which are contained in this publication.

Examples and articles contained within this book are works of fiction. Names, characters, businesses, places, events and incidents are either the products of the author's imagination or used in a fictitious manner. Any resemblance to actual persons, living or dead, or actual events is purely coincidental.

The publishers would like to thank the following who have given permission to reproduce photographs in this book:

Photo credits: page 1 © yuhorakushin/stock.adobe.com; page 2 © Artur Shevel/Fotolia; page 52 © Ingram Publishing Limited/General Gold Vol 1 CD 2; page 78 *top* © molekuul/123.com; page 78 *bottom* © Luminis/Fotolia; page 83 © Adul10/Shutterstock; page 108 © Ivan Kuzmin/Alamy; page 126 *top* © adike/Shutterstock; page 126 *bottom* © Emma Lee/Alamy Stock Photo; page 137 © Peter Küng/Fotolia; page 145 © Milles Studio/Shutterstock

Every effort has been made to trace and acknowledge ownership of copyright. The publishers will be glad to make suitable arrangements with any copyright holders whom it has not been possible to contact.

Hachette UK's policy is to use papers that are natural, renewable and recyclable products and made from wood grown in well-managed forests and other controlled sources. The logging and manufacturing processes are expected to conform to the environmental regulations of the country of origin.

Orders: please contact Hachette UK Distribution, Hely Hutchinson Centre, Milton Road, Didcot, Oxfordshire, OX11 7HH. Telephone: +44 (0)1235 827827. Email education@hachette.co.uk Lines are open from 9 a.m. to 5 p.m., Monday to Friday. You can also order through our website: www.hoddereducation.com

Much of the material in this book was published originally as part of the MEI Structured Mathematics series. It has been carefully adapted for the Cambridge International AS & A Level Mathematics syllabus. The original MEI author team for Statistics comprised Alec Cryer, Michael Davies, Anthony Eccles, Bob Francis, Gerald Goddall, Alan Graham, Nigel Green, Liam Hennessey, Roger Porkess and Charlie Stripp.

First published in 2018 by
Hodder Education,
an Hachette UK company,
Carmelite House,
50 Victoria Embankment,
London EC4Y 0DZ

Impression number   10 9

Year   2024

Cover photo © Shutterstock/Andrei Tudoran
Illustrations by Pantek Media, Maidstone, Kent & Integra Software Services Pvt Ltd, Pondicherry, India; page 145 *top* by Bibin Jose
Typeset in Bembo Std 11/13 by Integra Software Services Pvt Ltd, Pondicherry, India
Printed in India

A catalogue record for this title is available from the British Library.

ISBN 978 15104 2175 2

MIX
Paper | Supporting responsible forestry
FSC www.fsc.org FSC™ C104740

Contents

Introduction

This is one of a series of five books supporting the Cambridge International AS & A Level Mathematics 9709 syllabus for examination from 2020. The series then continues with four more books supporting Cambridge International AS & A Level Further Mathematics 9231. The seven chapters in this book cover the probability and statistics required for the Paper 5 examination. This part of the series also contains a more advanced book for probability and statistics, two books for pure mathematics and one book for mechanics.

These books are based on the highly successful series for the Mathematics in Education and Industry (MEI) syllabus in the UK but they have been redesigned and revised for Cambridge International students; where appropriate, new material has been written and the exercises contain many past Cambridge International examination questions. An overview of the units making up the Cambridge International syllabus is given in the following pages.

Throughout the series, the emphasis is on understanding the mathematics as well as routine calculations. The various exercises provide plenty of scope for practising basic techniques; they also contain many typical examination-style questions.

The original MEI author team would like to thank Sophie Goldie who has carried out the extensive task of presenting their work in a suitable form for Cambridge International students and for her many original contributions. They would also like to thank Cambridge Assessment International Education for its detailed advice in preparing the books and for permission to use many past examination questions.

Roger Porkess

Series editor

How to use this book

The structure of the book

This book has been endorsed by Cambridge Assessment International Education. It is listed as an endorsed textbook for students taking the Cambridge International AS & A Level Mathematics 9709 syllabus. The Probability & Statistics 1 syllabus content is covered comprehensively and is presented across seven chapters, offering a structured route through the course.

The book is written on the assumption that you have covered and understood the content of the Cambridge IGCSE™ Mathematics 0580 (Extended curriculum) or Cambridge O Level Mathematics 4024/4029 syllabus. The following icon is used to indicate material that is not directly on the syllabus.

(e) There are places where the book goes beyond the requirements of the syllabus to show how the ideas can be taken further or where fundamental underpinning work is explored. Such work is marked as **extension**.

Each chapter is broken down into several sections, with each section covering a single topic. Topics are introduced through **explanations**, with **key terms** picked out in red. These are reinforced with plentiful **worked examples**, punctuated with commentary, to demonstrate methods and illustrate application of the mathematics under discussion.

Regular **exercises** allow you to apply what you have learned. They offer a large variety of practice and higher-order question types that map to the key concepts of the Cambridge International syllabus. Look out for the following icons.

PS **Problem-solving questions** will help you to develop the ability to analyse problems, recognise how to represent different situations mathematically, identify and interpret relevant information, and select appropriate methods.

M **Modelling questions** provide you with an introduction to the important skill of mathematical modelling. In this, you take an everyday or workplace situation, or one that arises in your other subjects, and present it in a form that allows you to apply mathematics to it.

CP **Communication and proof questions** encourage you to become a more fluent mathematician, giving you scope to communicate your work with clear, logical arguments and to justify your results.

Exercises also include questions from real Cambridge Assessment International Education past papers, so that you can become familiar with the types of questions you are likely to meet in formal assessments.

Answers to exercise questions, excluding long explanations and proofs, are included in the back of the book, so you can check your work. It is important, however, that you have a go at answering the questions before looking up the answers if you are to understand the mathematics fully.

In addition to the exercises, a range of additional features is included to enhance your learning.

> ## ACTIVITY
>
> **Activities** invite you to do some work for yourself, typically to introduce you to ideas that are then going to be taken further. In some places, activities are also used to follow up work that has just been covered.

PROBLEM SOLVING　**PS**

Mathematics provides you with the techniques to answer many standard questions, but it also does much more than that: it helps you to develop the capacity to analyse problems and to decide for yourself what methods and techniques you will need to use. Questions and situations where this is particularly relevant are highlighted as **problem solving tasks**.

INVESTIGATION

In real life, it is often the case that as well as analysing a situation or problem, you also need to carry out some investigative work. This allows you to check whether your proposed approach is likely to be fruitful or to work at all, and whether it can be extended. Such opportunities are marked as **investigations**.

EXPERIMENT

In applied mathematics (mechanics and statistics), it is often helpful to carry out **experiments** so that you can see for yourself what is going on. The same is sometimes true for pure mathematics, where a spreadsheet can be a particularly powerful tool.

Other helpful features include the following.

? This symbol highlights points it will benefit you to **discuss** with your teacher or fellow students, to encourage deeper exploration and mathematical communication. If you are working on your own, there are answers in the back of the book.

! This is a **warning** sign. It is used where a common mistake, misunderstanding or tricky point is being described to prevent you from making the same error.

> ## Note
>
> **Notes** expand on the topic under consideration and explore the deeper lessons that emerge from what has just been done.

Finally, each chapter ends with the **key points** covered, plus a list of the **learning outcomes** that summarise what you have learned in a form that is closely related to the syllabus.

Digital support

Comprehensive online support for this book, including further questions, is available by subscription to MEI's Integral® online teaching and learning platform for AS & A Level Mathematics and Further Mathematics, integralmaths.org. This online platform provides extensive, high-quality resources, including printable materials, innovative interactive activities, and formative and summative assessments. Our eTextbooks link seamlessly with Integral, allowing you to move with ease between corresponding topics in the eTextbooks and Integral.

Additional support

The **Question & Workbooks** provide additional practice for students. These write-in workbooks are designed to be used throughout the course.

The **Study and Revision Guides** provide further practice for students as they prepare for their examinations.

These supporting resources and MEI's Integral® material have not been through the Cambridge International endorsement process.

The Cambridge International AS & A Level Mathematics 9709 syllabus

The syllabus content is assessed over six examination papers.

Paper 1: Pure Mathematics 1	Paper 4: Mechanics
• 1 hour 50 minutes	• 1 hour 15 minutes
• 60% of the AS Level; 30% of the A Level	• 40% of the AS Level; 20% of the A Level
• Compulsory for AS and A Level	• Offered as part of AS or A Level
Paper 2: Pure Mathematics 2	**Paper 5: Probability & Statistics 1**
• 1 hour 15 minutes	• 1 hour 15 minutes
• 40% of the AS Level	• 40% of the AS Level; 20% of the A Level
• Offered only as part of AS Level; not a route to A Level	• Compulsory for A Level
Paper 3: Pure Mathematics 3	**Paper 6: Probability & Statistics 2**
• 1 hour 50 minutes	• 1 hour 15 minutes
• 30% of the A Level	• 20% of the A Level
• Compulsory for A Level; not a route to AS Level	• Offered only as part of A Level; not a route to AS Level

The following diagram illustrates the permitted combinations for AS Level and A Level.

AS Level Mathematics **A Level Mathematics**

Paper 1 and Paper 2
Pure Mathematics only

(No progression to A Level)

Paper 1 and Paper 4
Pure Mathematics and Mechanics

Paper 1 and Paper 5
Pure Mathematics and
Probability & Statistics

Paper 1, 3, 4 and 5
Pure Mathematics, Mechanics and
Probability & Statistics

Paper 1, 3, 5 and 6
Pure Mathematics and
Probability & Statistics

Prior knowledge

Knowledge of the content of the Cambridge IGCSE™ Mathematics 0580 (Extended curriculum), or Cambridge O Level 4024/4029, is assumed. Learners should be familiar with scientific notation for compound units, e.g. $5\,\text{m\,s}^{-1}$ for 5 metres per second.

In addition, learners should:

» be able to carry out simple manipulation of surds (e.g. expressing $\sqrt{12}$ as $2\sqrt{3}$ and $\dfrac{6}{\sqrt{2}}$ as $3\sqrt{2}$)

» know the shapes of graphs of the form $y = kx^n$, where k is a constant and n is an integer (positive or negative) or $\pm\frac{1}{2}$.

Questions set will be mainly numerical, and will test principles in probability and statistics without involving knowledge of algebraic methods beyond the content for Paper 1: Pure Mathematics 1.

Knowledge of the following probability notation is also assumed: $P(A)$, $P(A \cup B)$, $P(A \cap B)$, $P(A \mid B)$ and the use of A' to denote the complement of A.

Command words

The table below includes command words used in the assessment for this syllabus. The use of the command word will relate to the subject context.

Command word	What it means
Calculate	work out from given facts, figures or information
Describe	state the points of a topic / give characteristics and main features
Determine	establish with certainty
Evaluate	judge or calculate the quality, importance, amount, or value of something
Explain	set out purposes or reasons / make the relationships between things evident / provide why and/or how and support with relevant evidence
Identify	name/select/recognise
Justify	support a case with evidence/argument
Show (that)	provide structured evidence that leads to a given result
Sketch	make a simple freehand drawing showing the key features
State	express in clear terms
Verify	confirm a given statement/result is true

Key concepts

Key concepts are essential ideas that help students develop a deep understanding of mathematics.

The key concepts are:

Problem solving

Mathematics is fundamentally problem solving and representing systems and models in different ways. These include:

» Algebra: this is an essential tool which supports and expresses mathematical reasoning and provides a means to generalise across a number of contexts.

» Geometrical techniques: algebraic representations also describe a spatial relationship, which gives us a new way to understand a situation.

» Calculus: this is a fundamental element which describes change in dynamic situations and underlines the links between functions and graphs.

» Mechanical models: these explain and predict how particles and objects move or remain stable under the influence of forces.

» Statistical methods: these are used to quantify and model aspects of the world around us. Probability theory predicts how chance events might proceed, and whether assumptions about chance are justified by evidence.

Communication

Mathematical proof and reasoning is expressed using algebra and notation so that others can follow each line of reasoning and confirm its completeness and accuracy. Mathematical notation is universal. Each solution is structured, but proof and problem solving also invite creative and original thinking.

Mathematical modelling

Mathematical modelling can be applied to many different situations and problems, leading to predictions and solutions. A variety of mathematical content areas and techniques may be required to create the model. Once the model has been created and applied, the results can be interpreted to give predictions and information about the real world.

These key concepts are reinforced in the different question types included in this book: **Problem-solving**, **Communication and proof**, and **Modelling**.

1

Exploring data

A judicious man looks at statistics, not to get knowledge but to save himself from having ignorance foisted on him.
Thomas Carlyle (1795–1881)

> A Sales Executive has been given the data shown above in order to write her company's annual Sales and Marketing report. Some of the data are shown as figures and some as diagrams.
> ❯ How should she interpret this information?
> ❯ What data should she take seriously and which can she dismiss as being insignificant or even misleading?

To answer these questions fully you need to understand how data are collected and analysed before they are presented to you, and how you should evaluate what you are given to read (or see on the television). This is an important part of the subject of statistics.

In this book, many of the examples are set as stories from fictional websites. Some of them are written as articles or blogs; others are presented from the journalists' viewpoint as they sort through data trying to write an interesting story. As you work through the book, look too at the ways you are given such information in your everyday life.

Biking Today Blog

Another cyclist seriously hurt. *Will you be next?*

On her way back home from school on Wednesday afternoon, little Rita Roy was knocked off her bicycle and taken to hospital with suspected concussion.

Rita was struck by a Ford Transit van, only 50 metres from her own house.

Rita is the fourth child from the Nelson Mandela estate to be involved in a serious cycling accident this year.

The busy road where Rita Roy was knocked off her bicycle yesterday.

After reading the blog, the editor of a local newspaper commissioned one of the paper's reporters to investigate the situation and write a leading article for the paper on it. She explained to the reporter that there was growing concern locally about cycling accidents involving children. She emphasised the need to collect good quality data to support presentations to the paper's readers.

> Is the aim of the investigation clear?
> Is the investigation worth carrying out?
> What makes good quality data?

The reporter started by collecting data from two sources. He went through back numbers of the newspaper for the previous two years, finding all the reports of cycling accidents. He also asked an assistant to carry out a survey of the ages of local cyclists; he wanted to know whether most cyclists were children, young adults or whatever.

> Are the reporter's data sources appropriate?

Before starting to write his article, the reporter needed to make sense of the data for himself. He then had to decide how he was going to present the information to his readers. The five examples on the page opposite are typical of the data he had to work with.

Name	Age	Distance from home	Cause	Injuries	Treatment
Rahim Khan	45	3 km	skid	Concussion	Hospital outpatient
Debbie Lane	5	75 m	hit kerb	Broken arm	Hospital outpatient
Arvinder Sethi	12	1200 m	lorry	Multiple fractures	Hospital 3 weeks
Husna Mahar	8	300 m	hit each other	Bruising	Hospital outpatient
David Huker	8	50 m		Concussion	Hospital outpatient

There were 92 accidents listed in his table.

Ages of cyclists (from survey)

```
66   6  62   19  20      15  21   8  21  63      44  10  44  34  18
35  26  61   13  61      28  21   7  10  52      13  52  20  17  26
64  11  39   22   9      13   9  17  64  32       8   9  31  19  22
37  18 138  16  67      45  10  55  14  66      67  14  62  28  36
 9  23  12    9  37       7  36   9  88  46      12  59  61  22  49
18  20  11   25   7      42  29   6  60  60      16  50  16  34  14
18  15
```

This information is described as **raw data**, which means that no attempt has yet been made to organise it to look for any patterns.

1.1 Looking at the data

At the moment the arrangement of the ages of the 92 cyclists tells you very little at all. Clearly these data must be organised to reveal the underlying shape, the **distribution**. The figures need to be ranked according to size and preferably grouped as well. The reporter had asked an assistant to collect the information and this was the order in which she presented it.

Tally

Tallying is a quick, straightforward way of grouping data into suitable intervals. You have probably met it already.

Stated age (years)	Tally	Frequency
0–9	ЖЖ ЖЖ |||	13
10–19	ЖЖ ЖЖ ЖЖ ЖЖ ЖЖ		26
20–29	ЖЖ ЖЖ ЖЖ		16
30–39	ЖЖ ЖЖ	10
40–49	ЖЖ		6
50–59	ЖЖ	5
60–69	ЖЖ ЖЖ ||||	14
70–79		0
80–89			1
⋮		
130–139			1
Total		**92**

Extreme values

A tally immediately shows up any extreme values, that is values which are far away from the rest. In this case there are two extreme values, usually referred to as **outliers**: 88 and 138. Before doing anything else you must investigate these.

In this case the 88 is genuine, the age of Millie Smith, who is a familiar sight cycling to the shops.

The 138 needless to say is not genuine. It was the written response of a man who was insulted at being asked his age. Since no other information about him is available, this figure is best ignored and the sample size reduced from 92 to 91. You should always try to understand an outlier before deciding to ignore it; it may give you important information.

> Practical statisticians are frequently faced with the problem of **outlying observations**; observations that depart in some way from the general pattern of a data set. What they, and you, have to decide is whether any such observations belong to the data set or not. In the above example the data value 88 is a genuine member of the data set and is retained. The data value 138 is not a member of the data set and is therefore rejected.

Describing the shape of a distribution

An obvious benefit of using a tally is that it shows the overall shape of the distribution.

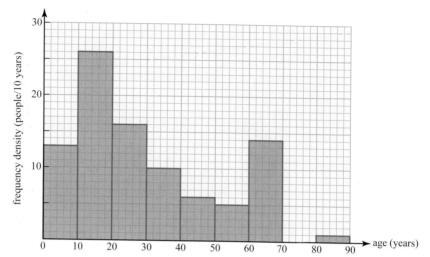

▲ **Figure 1.1** Histogram to show the ages of a sample of 91 cyclists

You can now see that a large proportion (more than a quarter) of the sample are in the 10 to 19 years age range. This is the **modal** group as it is the one with the most members. The single value with the most members is called the **mode**, in this case age 9.

You will also see that there is a second peak among those in their sixties; so this distribution is called **bimodal**, even though the frequency in the interval 10–19 is greater than the frequency in the interval 60–69.

Different types of distribution are described in terms of the position of their modes or modal groups; see Figure 1.2.

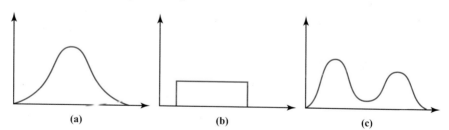

▲ **Figure 1.2** Distribution shapes: (a) unimodal and symmetrical (b) uniform (no mode but symmetrical) (c) bimodal

When the mode is off to one side the distribution is said to be **skewed**. If the mode is to the left with a long tail to the right the distribution has positive (or right) skewness; if the long tail is to the left the distribution has negative (or left) skewness. These two cases are shown in Figure 1.3.

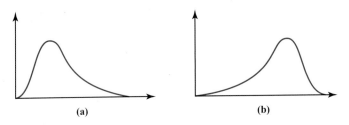

▲ **Figure 1.3** Skewness: (a) positive (b) negative

1.2 Stem-and-leaf diagrams

The quick and easy view of the distribution from the tally has been achieved at the cost of losing information. You can no longer see the original figures that went into the various groups and so cannot, for example, tell from looking at the tally whether Millie Smith is 80, 81, 82, or any age up to 89. This problem of the loss of information can be solved by drawing a **stem–and–leaf diagram** (or **stemplot**).

This is a quick way of grouping the data so that you can see their distribution and still have access to the original figures. The one below shows the ages of the 91 cyclists surveyed.

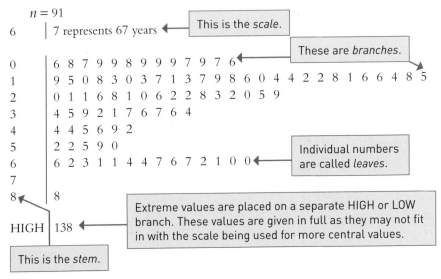

▲ **Figure 1.4** Stem-and-leaf diagram showing the ages of a sample of 91 cyclists (unsorted)

> ❯ Do all the branches have leaves?

The column of figures on the left (going from 0 to 8) corresponds to the tens digits of the ages. This is called the **stem** and in this example it consists of 9 branches. On each branch on the stem are the **leaves** and these represent the units digits of the data values.

In Figure 1.4, the leaves for a particular branch have been placed in the order in which the numbers appeared in the original raw data. This is fine for showing the general shape of the distribution, but it is usually worthwhile sorting the leaves, as shown in Figure 1.5.

$n = 91$

6 | 7 represents 67 years

0	6 6 7 7 7 8 8 9 9 9 9 9 9
1	0 0 0 1 1 2 2 3 3 3 4 4 4 5 5 6 6 6 7 7 8 8 8 8 9 9
2	0 0 0 1 1 1 2 2 2 3 5 6 6 8 8 9
3	1 2 4 4 5 6 6 7 7 9
4	2 4 4 5 6 9
5	0 2 2 5 9
6	0 0 1 1 1 2 2 3 4 4 6 6 7 7
7	
8	8

Note that the value 138 is left out as it has been identified as not belonging to this set of data.

▲ **Figure 1.5** Stem-and-leaf diagram showing the ages of a sample of 91 cyclists (sorted)

The stem-and-leaf diagram gives you a lot of information at a glance:

➤ The youngest cyclist is 6 and the oldest is 88 years of age.

➤ More people are in the 10–19 years age range than in any other 10-year age range.

➤ There are three 61 year olds.

➤ The modal age (i.e. the age with the most people) is 9.

➤ The 17th oldest cyclist in the survey is 55 years of age.

If the values on the basic stem-and-leaf diagram are too cramped, that is if there are so many leaves on a line that the diagram is not clear, you may **stretch** it. To do this you put values 0, 1, 2, 3, 4 on one line and 5, 6, 7, 8, 9 on another. Doing this to the example results in the diagram shown in Figure 1.6.

When stretched, this stem-and-leaf diagram reveals the skewed nature of the distribution.

$n = 91$

6 | 7 represents 67 years

0★														
0		6	6	7	7	7	8	8	9	9	9	9	9	9
1★		0	0	0	1	1	2	2	3	3	3	4	4	4
1		5	5	6	6	6	7	7	8	8	8	8	9	9
2★		0	0	0	1	1	1	2	2	2	3			
2		5	6	6	8	8	9							
3★		1	2	4	4									
3		5	6	6	7	7	9							
4★		2	4	4										
4		5	6	9										
5★		0	2	2										
5		5	9											
6★		0	0	1	1	1	2	2	3	4	4			
6		6	6	7	7									
7★														
7														
8★														
8		8												

▲ Figure 1.6 Stem-and-leaf diagram showing the ages of a sample of 91 cyclists (sorted, stretched)

> ❯ How would you squeeze a stem-and-leaf diagram?
> ❯ What would you do if the data have more significant figures than can be shown on a stem-and-leaf diagram?

Stem-and-leaf diagrams are particularly useful for comparing data sets. With two data sets a back-to-back stem-and-leaf diagram can be drawn, as shown in Figure 1.7.

represents 590   9 | 5 | 2   represents 520

```
            9 | 5 | 1 7
            2 | 6 | 0 2 3 5 8
        5 3 0 | 7 | 1 2 5 6 6 7
    9 7 5 1 1 | 8 | 3 5
      8 6 2 1 | 9 | 2
```

▲ Figure 1.7

Note the numbers on the left of the stem still have the smallest number next to the stem.

> ❯ How would you represent positive and negative data on a stem-and-leaf diagram?

1   Write down the lengths in centimetres that are represented by this stem-and-leaf diagram.

$n = 15$
32 | 1   represents 3.21 cm

32	7
33	2 6
34	3 5 9
35	0 2 6 6 8
36	1 1 4
37	2

2   Write down the lengths in millimetres that are represented by this stem-and-leaf diagram.

$n = 19$
8 | 9   represents 0.089 mm

8	3 6 7
9	0 1 4 8
10	2 3 5 8 9 9
11	0 1 4
12	3 5
13	1

3   Draw a sorted stem-and-leaf diagram with six branches to show the following numbers. Remember to include the appropriate scale.

0.212   0.223   0.226   0.230   0.233   0.237   0.241

0.242   0.248   0.253   0.253   0.259   0.262

4   Draw a sorted stem-and-leaf diagram with five branches to show the following numbers. Remember to include the appropriate scale.

81.7   82.0   78.1   80.8   82.5

81.9   79.4   81.3   79.6   80.4

5 Write down the lengths in metres that are represented by this stem-and-leaf diagram.

$n = 21$

34 | 5   represents 3.45 m

LOW   0.013, 0.089, 1.79

34	3
35	1 7 9
36	0 4 6 8
37	1 1 3 8 9
38	0 5
39	4

HIGH   7.45, 10.87

6 Forty motorists entered a driving competition. The organisers were anxious to know if the contestants had enjoyed the event and also to know their ages, so that they could plan and promote future events effectively. They therefore asked entrants to fill in a form on which they commented on the various tests and gave their ages.

The information was copied from the forms and the ages listed as:

28	52	44	28	38		46	62	59	37	60
19	55	34	35	66		37	22	26	45	5
61	38	26	29	63		38	29	36	45	33
37	41	39	81	35		35	32	36	39	33

(i)   Identify the outlier and say why it should be excluded.

(ii)   Draw a sorted stem-and-leaf diagram to illustrate these data.

(iii)   Describe the shape of the distribution.

CP

7 The unsorted stem-and-leaf diagram below gives the ages of males whose marriages were reported in a local newspaper one week.

$n = 42$

1 | 9   represents 19

0	
1	9 6 9 8
2	5 6 8 9 1 1 0 3 6 8 4 1 2 7
3	0 0 5 2 3 9 1 2 0
4	8 4 7 9 6 5 3 3 5 6
5	2 2 1 7
6	
7	
8	3

(i) What was the age of the oldest person whose marriage is included?

(ii) Redraw the stem-and-leaf diagram with the leaves sorted.

(iii) Stretch the stem-and-leaf diagram by using steps of five years between the levels rather than ten.

(iv) Describe and comment on the distribution.

8 On 1 January the average daily temperature was recorded for 30 cities around the world. The temperatures, in °C, were as follows.

21	3	18	−4	10		27	14	7	19	−14
32	2	−9	29	11		26	−7	−11	15	4
35	14	23	19	−15		8	8	−2	3	1

(i) Draw a sorted stem-and-leaf diagram to illustrate the temperature distribution.

(ii) Describe the shape of the distribution.

CP 9 The following marks were obtained on an A Level mathematics paper by the candidates at one centre.

26	54	50	37	54		34	34	66	44	76		45	71	51	75	30
29	52	43	66	59		22	74	51	49	39		32	37	57	37	18
54	17	26	40	69		80	90	95	96	95		70	68	97	87	68
77	76	30	100	98		44	60	46	97	75		52	82	92	51	44
73	87	49	90	53		45	40	61	66	94		62	39	100	91	66
35	56	36	74	25		70	69	67	48	65		55	64			

Draw a sorted stem-and-leaf diagram to illustrate these marks and comment on their distribution.

CP 10 The ages of a sample of 40 hang-gliders (in years) are given below.

28	19	24	20	28		26	22	19	37	40		19	25	65	34	66
35	69	65	26	17		22	26	45	58	30		31	58	26	29	23
72	23	21	30	28		65	21	67	23	57						

(i) Using intervals of ten years, draw a sorted stem-and-leaf diagram to illustrate these figures.

(ii) Comment on and give a possible explanation for the shape of the distribution.

CP **11** An experimental fertiliser called GRO was applied to 50 lime trees, chosen at random, in a plantation. Another 50 trees were left untreated. The yields in kilograms were as follows.

Treated

59	25	52	19	32		26	33	24	35	30		23	54	33	31	25
23	61	35	38	44		27	24	30	62	23		47	42	41	53	31
20	21	41	33	35		38	61	63	44	18		53	38	33	49	54
50	44	25	42	18												

Untreated

8	11	22	22	20		5	31	40	14	45		10	16	14	20	51
55	30	30	25	29		12	48	17	12	52		58	61	14	32	5
29	40	61	53	22		33	41	62	51	56		10	48	50	14	8
63	43	61	12	42												

Draw sorted back-to-back stem-and-leaf diagrams to compare the two sets of data and comment on the effects of GRO.

CP **12** A group of 25 people were asked to estimate the length of a line which they were told was between 1 and 2 metres long. Here are their estimates, in metres.

$$1.15 \quad 1.33 \quad 1.42 \quad 1.26 \quad 1.29 \qquad 1.30 \quad 1.30 \quad 1.46 \quad 1.18 \quad 1.24$$

$$1.21 \quad 1.30 \quad 1.32 \quad 1.33 \quad 1.29 \qquad 1.30 \quad 1.40 \quad 1.26 \quad 1.32 \quad 1.30$$

$$1.41 \quad 1.28 \quad 1.65 \quad 1.54 \quad 1.14$$

(i) Represent these data in a sorted stem-and-leaf diagram.

(ii) From the stem-and-leaf diagram which you drew, read off the third highest and third lowest length estimates.

(iii) Find the middle of the 25 estimates.

(iv) On the evidence that you have, could you make an estimate of the length of the line? Justify your answer.

1.3 Categorical or qualitative data

Chapter 2 will deal in more detail with ways of displaying data. The remainder of this chapter looks at types of data and the basic analysis of numerical data.

Some data come to you in classes or categories. Such data, like these for the sizes of sweatshirts, are called categorical or qualitative.

XL, S, S, L, M, S, M, M, XL, L, XS
XS = extra small; S = small; M = medium; L = large; XL = extra large

Most of the data you encounter, however, will be numerical data (also called quantitative data).

1.4 Numerical or quantitative data

Variables

The score you get when you throw an ordinary die is one of the values 1, 2, 3, 4, 5 or 6. Rather than repeatedly using the phrase 'The score you get when you throw an ordinary die', statisticians find it convenient to use a capital letter, X, say. They let X stand for 'The score you get when you throw an ordinary die' and because this varies, X is referred to as a **variable**.

Similarly, if you are collecting data and this involves, for example, noting the temperature in classrooms at noon, then you could let T stand for 'the temperature in a classroom at noon'. So T is another example of a variable.

Values of the variable X are denoted by the lower case letter x, e.g. $x = 1, 2, 3, 4, 5$ or 6.

Values of the variable T are denoted by the lower case letter t, e.g. $t = 18, 21, 20, 19, 23, \ldots$.

Discrete and continuous variables

The scores on a die, 1, 2, 3, 4, 5 and 6, the number of goals a football team scores, 0, 1, 2, 3, ... and amounts of money, \$0.01, \$0.02, ... are all examples of **discrete variables**. What they have in common is that all possible values can be listed.

Distance, mass, temperature and speed are all examples of **continuous variables.** Continuous variables, if measured accurately enough, can take any appropriate value. You cannot list all possible values.

You have already seen the example of age. This is rather a special case. It is nearly always given rounded down (i.e. truncated). Although your age changes continuously every moment of your life, you actually state it in steps of one year, in completed years, and not to the nearest whole year. So a man who is a few days short of his 20th birthday will still say he is 19.

In practice, data for a continuous variable are always given in a rounded form.

» A person's height, h, given as 168 cm, measured to the nearest centimetre; $167.5 \leqslant h < 168.5$

» A temperature, t, given as 21.8 °C, measured to the nearest tenth of a degree; $21.75 \leqslant t < 21.85$

» The depth of an ocean, d, given as 9200 m, measured to the nearest 100 m; $9150 \leqslant d < 9250$

Notice the rounding convention here: if a figure is on the borderline it is rounded up. There are other rounding conventions.

1.5 Measures of central tendency

When describing a typical value to represent a data set most people think of a value at the centre and use the word **average**. When using the word average they are often referring to the **arithmetic mean**, which is usually just called the **mean**, and when asked to explain how to get the mean most people respond by saying 'add up the data values and divide by the total number of data values'.

There are actually several different averages and so, in statistics, it is important for you to be more precise about the *average* to which you are referring. Before looking at the different types of average or **measure of central tendency**, you need to be familiar with some notation.

Σ notation and the mean, $\bar{x}$

A sample of size n taken from a population can be identified as follows.

The first item can be called x_1, the second item x_2 and so on up to x_n.

The sum of these n items of data is given by $x_1 + x_2 + x_3 + ... + x_n$.

A shorthand for this is $\sum_{i=1}^{i=n} x_i$ or $\sum_{i=1}^{n} x_i$. This is read as 'the sum of all the terms x_i when i equals 1 to n'.

So $$\sum_{i=1}^{n} x_i = x_1 + x_2 + x_3 + ... + x_n.$$ ← Σ is the Greek capital letter, sigma.

If there is no ambiguity about the number of items of data, the subscripts i can be dropped and $\sum_{i=1}^{n} x_i$ becomes $\sum x$.

$\sum x$ is read as 'sigma x' meaning 'the sum of all the x items'.

The mean of these n items of data is written as $\bar{x} = \dfrac{x_1 + x_2 + x_3 + ... + x_n}{n}$

where $\bar{x}$ is the symbol for the mean, referred to as '*x-bar*'.

It is usual to write $\bar{x} = \dfrac{\sum x}{n}$ or $\dfrac{1}{n}\sum x$.

This is a formal way of writing 'To get the mean you add up all the data values and divide by the total number of data values'.

The mean from a frequency table

Often data is presented in a frequency table. The notation for the mean is slightly different in such cases.

Alex is a member of the local bird-watching group. The group are concerned about the effect of pollution and climatic change on the well-being of birds. One spring Alex surveyed the nests of a type of owl. Healthy owls usually lay up to 6 eggs. Alex collected data from 50 nests. His data are shown in the following frequency table.

Number of eggs, x	Frequency, f
1	4
2	12
3	9
4	18
5	7
6	0
Total	$\Sigma f = 50$

> This represents 'the sum of the separate frequencies is 50'. That is, 4 + 12 + 9 + 18 + 7 = 50

It would be possible to write out the data set in full as 1, 1, 1, … , 5, 5 and then calculate the mean as before. However, it would not be sensible and in practice the mean is calculated as follows:

$$\bar{x} = \frac{1 \times 4 + 2 \times 12 + 3 \times 9 + 4 \times 18 + 5 \times 7}{50}$$

$$= \frac{162}{50}$$

$$= 3.24$$

In general, this is written as $\bar{x} = \dfrac{\sum xf}{n}$

> This represents the sum of each of the x terms multiplied by its frequency.

$n = \Sigma f$

In the survey at the beginning of this chapter the mean of the cyclists' ages,

$$\bar{x} = \frac{2717}{91} = 29.9 \text{ years.}$$

However, a mean of the ages needs to be adjusted because age is always rounded down. For example, Rahim Khan gave his age as 45. He could be exactly 45 years old or possibly his 46th birthday may be one day away. So, each of the people in the sample could be from 0 to almost a year older than their quoted age. To adjust for this discrepancy you need to add 0.5 years on to the average of 29.9 to give 30.4 years.

Note

The mean is the most commonly used average in statistics. The mean described here is correctly called the *arithmetic mean*; there are other forms, for example the geometric mean, harmonic mean and weighted mean, all of which have particular applications.

The mean is used when the total quantity is also of interest. For example, the staff at the water treatment works for a city would be interested in the mean amount of water used per household ($\bar{x}$) but would also need to know the total amount of water used in the city (Σx). The mean can give a misleading result if exceptionally large or exceptionally small values occur in the data set.

There are two other commonly used statistical measures of a typical (or representative) value of a data set. These are the median and the mode.

Median

The median is the value of the middle item when all the data items are ranked in order. If there are n items of data then the median is the value of the $\frac{n+1}{2}$th item.

If n is odd then there is a middle value and this is the median. In the survey of the cyclists we have

> The 46th item of data is 22 years.

$$6, 6, 7, 7, 7, 8, \ldots, 20, 21, 21, 21, 22, 22, 22, \ldots$$

So for the ages of the 91 cyclists, the median is the age of the $\frac{91+1}{2} = 46$th person and this is 22 years.

If n is even and the two middle values are a and b then the median is $\frac{a+b}{2}$.

For example, if the reporter had not noticed that 138 was invalid there would have been 92 items of data. Then the median age for the cyclists would be found as follows.

> The 46th and 47th items of data are the two middle values and are both 22.

$$6, 6, 7, 7, 7, 8, \ldots, 20, 21, 21, 21, 22, 22, 22, \ldots$$

So the median age for the cyclists is given as the mean of the 46th and 47th items of data. That is, $\frac{22+22}{2} = 22$.

It is a coincidence that the median turns out to be the same. However, what is important to notice is that an extreme value has little or no effect on the value of the median. The median is said to be resistant to outliers.

The **median** is easy to work out if the data are already ranked, otherwise it can be tedious. However, with the increased availability of computers, it is easier to sort data and so the use of the median is increasing. Illustrating data on a stem-and-leaf diagram orders the data and makes it easy to identify the median. The median usually provides a good representative value and, as seen above, it is not affected by extreme values. It is particularly useful if some values are missing; for example, if 50 people took part in a marathon then the median is halfway between the 25th and 26th values. If some people failed to complete the course the mean would be impossible to calculate, but the median is easy to find.

In finding an **average** salary the median is often a more appropriate measure than the mean since a few people earning very large salaries may have a big effect on the mean but not on the median.

Mode

The **mode** is the value which occurs most frequently. If two non-adjacent values occur more frequently than the rest, the distribution is said to be **bimodal**, even if the frequencies are not the same for both modes.

Bimodal data usually indicates that the sample has been taken from two populations. For example, a sample of students' heights (male and female) would probably be bimodal reflecting the different average heights of males and females.

For the cyclists' ages, the mode is 9 years (the frequency is 6).

For a small set of discrete data the mode can often be misleading, especially if there are many values the data can take. Several items of data can happen to fall on a particular value. The mode is used when the most probable or most frequently occurring value is of interest. For example, a dress shop manager who is considering stocking a new style would first buy dresses of the new style in the modal size, as she would be most likely to sell those ones.

Which average you use will depend on the particular data you have and on what you are trying to find out.

The measures for the cyclists' ages are summarised below.

> Mean   29.9 years   (adjusted = 30.4 years)
> Mode   9 years
> Median 22 years

> ❯ Which do you think is most representative?

Example 1.1

These are the times, in minutes, that a group of people took to answer a Sudoku puzzle.

> 5, 4, 11, 8, 4, 43, 10, 7, 12

Calculate an appropriate measure of central tendency to summarise these times. Explain why the other measures are not suitable.

Solution

First order the data.

> 4, 4, 5, 7, 8, 10, 11, 12, 43

➡

One person took much longer to solve the puzzle than the others, so the mean is not appropriate to use as it is affected by outliers.

The mode is 4, which is the lowest data value and is not representative of the data set.

So the most appropriate measure to use is the median.

There are nine data values; the median is the $\left(\dfrac{9+1}{2}\right)$th value, which is 8 minutes.

Exercise 1B

1   Find the mode, mean and median of these figures.

(i)   23 46 45 45 29   51 36 41 37 47   45 44 41 31 33

(ii)   110 111 116 119 129   126 132 116 122 130

116 132 118 122 127   132 126 138 117 111

(iii)   5 7 7 9 1   2 3 5 6 6   8 6 5 7 9   2 2 5 6 6

6 4 7 7 6   1 3 3 5 7   8 2 8 7 6   5 4 3 6 7

CP   2   For each of these sets of data:

(a)   find the mode, mean and median

(b)   state, with reasons, which you consider to be the most appropriate form of average to describe the distribution.

(i)   The ages of students in a class in years and months.

14.1 14.11 14.5 14.6 14.0   14.7 14.7 14.9 14.1   14.2

14.6 14.5   14.8 14.2 14.0   14.9 14.2 14.8 14.11   14.8

15.0 14.7   14.8 14.9 14.3   14.5 14.4 14.3 14.6   14.1

(ii)   Students' marks on an examination paper.

55 78 45 54   0 62 43 56 71 65   0 67 67 75 51 100

39 45 66 71   52 71   0   0 59 61   56 59 59 64 57   63

(iii)   The scores of a cricketer during a season's matches.

10 23 65   0   1   24 47   2 21 53   5   4   23 169 21

17 34 33 21   0   10 78   1 56   3   2   0 128   12 19

(iv)   Scores when a die is thrown 40 times.

2 4 5 5 1 3 4 6 2 5 2 4 6 1 2 5 4 4 1 1

3 4 6 5 5 2 3 3 1 6 5 4 2 1 3 3 2 1 6 6

3   The lengths of time in minutes to swim a certain distance by the members of a class of twelve 9-year-olds and by the members of a class of eight 16-year-olds are shown below.

9-year-olds: 13.0 16.1 16.0 14.4 15.9 15.1 14.2 13.7 16.7 16.4 15.0 13.2
16-year-olds: 14.8 13.0 11.4 11.7 16.5 13.7 12.8 12.9

(i) Draw a back-to-back stem-and-leaf diagram to represent the information above.

(ii) A new pupil joined the 16-year-old class and swam the distance. The mean time for the class of nine pupils was now 13.6 minutes. Find the new pupil's time to swim the distance.

Cambridge International AS & A Level Mathematics
9709 Paper 6 Q4 June 2007

1.6 Frequency distributions

You will often have to deal with data that are presented in a frequency table. Frequency tables summarise the data and also allow you to get an idea of the shape of the distribution.

Example 1.2

Claire runs a fairground stall. She has designed a game where customers pay $1 and are given 10 marbles which they have to try to get into a container 4 metres away. If they get more than 8 in the container they win $5. Before introducing the game to the customers she tries it out on a sample of 50 people. The number of successes scored by each person is noted.

5	7	8	7	5	4	0	9	10	6
4	8	8	9	5	6	3	2	4	4
6	5	5	7	6	7	5	6	9	2
7	7	6	3	5	5	6	9	8	7
5	2	1	6	8	5	4	4	3	3

> The data are discrete. They have not been organised in any way, so they are referred to as raw data.

Calculate the mode, median and mean scores. Comment on your results.

Solution

The **frequency distribution** of these data can be illustrated in a table. The number of 0s, 1s, 2s, etc. is counted to give the frequency of each mark.

Score	Frequency
0	1
1	1
2	3
3	4
4	6
5	10
6	8
7	7
8	5
9	4
10	1
Total	**50**

> With the data presented in this form it is easier to find or calculate the different averages.

→

The mode is 5 (frequency 10).

As the number of items of data is even, the distribution has two middle values, the 25th and 26th scores. From the distribution, by adding up the frequencies, it can be seen that the 25th score is 5 and the 26th score is 6. Consequently the median score is $\frac{1}{2}(5 + 6) = 5.5$.

Representing a score by x and its frequency by f, the calculation of the mean is shown below.

Score, x	Frequency, f	$x \times f$
0	1	$0 \times 1 = 0$
1	1	$1 \times 1 = 1$
2	3	$2 \times 3 = 6$
3	4	12
4	6	24
5	10	50
6	8	48
7	7	49
8	5	40
9	4	36
10	1	10
Totals	**50**	**276**

So $\bar{x} = \dfrac{\sum xf}{n}$ $\boxed{n = \Sigma f}$

$$= \frac{276}{50} = 5.52$$

The values of the mode (5), the median (5.5) and the mean (5.52) are close. This is because the distribution of scores does not have any extreme values and is reasonably symmetrical.

Example 1.3

The table shows the number of mobile phones owned by h households.

Number of mobile phones	0	1	2	3	4	5
Frequency	3	5	b	10	13	7

The mean number of mobile phones is 3. Find the values of b and h.

Solution

The total number of households, $h = 3 + 5 + b + 10 + 13 + 7$.

So $h = b + 38$

The total number of mobile phones $= 0 \times 3 + 1 \times 5 + 2 \times b + 3 \times 10 + 4 \times 13 + 5 \times 7$

$$= 2b + 122$$

$$\text{Mean} = \frac{\text{total number of mobile phones}}{\text{total frequency}} = 3$$

So

$$\frac{2b + 122}{b + 38} = 3$$

$$2b + 122 = 3(b + 38)$$

$$2b + 122 = 3b + 114$$

So $b = 8$ and $h = 8 + 38 = 46$.

Exercise 1C

1 A bag contained six counters numbered 1, 2, 3, 4, 5 and 6. A counter was drawn from the bag, its number was noted and then it was returned to the bag. This was repeated 100 times. The results were recorded in a table giving the frequency distribution shown.

(i) State the mode.

(ii) Find the median.

(iii) Calculate the mean.

Number, x	Frequency, f
1	15
2	25
3	16
4	20
5	13
6	11

CP

2 A sample of 50 boxes of matches with stated contents 40 matches was taken. The actual number of matches in each box was recorded. The resulting frequency distribution is shown in the table.

Number of matches, x	Frequency, f
37	5
38	5
39	10
40	8
41	7
42	6
43	5
44	4

(i) State the mode.

(ii) Find the median.

(iii) Calculate the mean.

(iv) State, with reasons, which you think is the most appropriate form of average to describe the distribution.

CP **3** A survey of the number of students in 80 classrooms in Avonford College was carried out. The data were recorded in a table as follows.

(i) State the mode.

(ii) Find the median.

(iii) Calculate the mean.

(iv) State, with reasons, which you think is the most appropriate form of average to describe the distribution.

Number of students, x	Frequency, f
5	1
11	1
15	6
16	9
17	12
18	16
19	18
20	13
21	3
22	1
Total	**80**

CP **4** The tally below gives the scores of the football teams in the matches of the 1982 World Cup finals.

Score	Tally
0	⊞⊞⊞ ⊞⊞⊞ ⊞⊞⊞ ⊞⊞⊞ ⊞⊞⊞ ⊞⊞⊞ |
1	⊞⊞⊞ ⊞⊞⊞ ⊞⊞⊞ ⊞⊞⊞ ⊞⊞⊞ ⊞⊞⊞ ⊞⊞⊞ ⫼
2	⊞⊞⊞ ⊞⊞⊞ ⊞⊞⊞ |
3	⊞⊞⊞ ⫼
4	⊞⊞⊞ |
5	⫼
6	
7	
8	
9	
10	|

(i) Find the mode, mean and median of these data.

(ii) State which of these you think is the most representative measure.

(For football enthusiasts: find out which team conceded 10 goals and why.)

CP **5** The vertical line chart below shows the number of times the various members of a school year had to take their driving test before passing it.

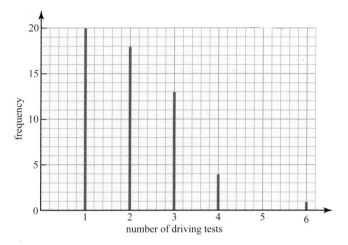

number of driving tests

(i) Find the mode, mean and median of these data.

(ii) State which of these you think is the most representative measure.

1.7 Grouped data

Grouping means putting the data into a number of classes. The number of data items falling into any class is called the **frequency** for that class.

When numerical data are grouped, each item of data falls within a **class interval** lying between **class boundaries**.

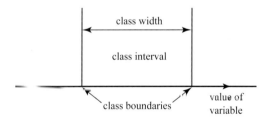

▲ **Figure 1.8**

You must always be careful about the choice of class boundaries because it must be absolutely clear to which class any item belongs. A form with the following wording:

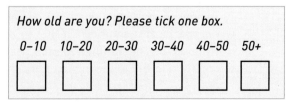

would cause problems. A ten-year-old could tick either of the first two boxes.

A better form of wording would be:

> **How old are you (in completed years)? Please tick one box.**
>
> 0–9   10–19   20–29   30–39   40–49   50+
>
> ☐   ☐   ☐   ☐   ☐   ☐

Notice that this says 'in completed years'. Otherwise a $9\frac{1}{2}$-year-old might not know which of the first two boxes to tick.

Another way of writing this is:

$0 \leqslant A < 10$	$10 \leqslant A < 20$	$20 \leqslant A < 30$
$30 \leqslant A < 40$	$40 \leqslant A < 50$	$50 \leqslant A$

Even somebody aged 9 years and 364 days would clearly still come in the first group.

> **?**
>
> Another way of writing these classes, which you will sometimes see, is:
> $$0–, 10–, 20–, \dots, 50–.$$
> ❯ What is the disadvantage of this way?

Working with grouped data

There is often a good reason for grouping raw data.

» There may be a lot of data.

» The data may be spread over a wide range.

» Most of the values collected may be different.

Whatever the reason, grouping data should make it easier to analyse and present a summary of findings, whether in a table or in a diagram.

For some **discrete data** it may not be necessary or desirable to group them. For example, a survey of the number of passengers in cars using a busy road is unlikely to produce many integer values outside the range 0 to 4 (not counting the driver). However, there are cases when grouping the data (or perhaps drawing a stem-and-leaf diagram) is an advantage.

Discrete data

At various times during one week the number of cars passing a survey point was noted. Each item of data relates to the number of cars passing during a five-minute period. A hundred such periods were surveyed. The data is summarised in the following frequency table.

Number of cars, x	Frequency, f
0–9	5
10–19	8
20–29	13
30–39	20
40–49	22
50–59	21
60–70	11
Total	**100**

From the frequency table you can see there is a slight negative (or left) skew.

Estimating the mean

When data are grouped the individual values are lost. This is not often a serious problem; as long as the data are reasonably distributed throughout each interval it is possible to **estimate** statistics such as the mean, knowing that your answers will be reasonably accurate.

To estimate the mean you first assume that all the values in an interval are equally spaced about a midpoint. The midpoints are taken as representative values of the intervals.

The mid-value for the interval 0–9 is $\dfrac{0+9}{2} = 4.5$.

The mid-value for the interval 10–19 is $\dfrac{10+19}{2} = 14.5$, and so on.

The $x \times f$ column can now be added to the frequency distribution table and an estimate for the mean found.

Number of cars, x (mid-values)	Frequency, f	$x \times f$
4.5	5	$4.5 \times 5 = 22.5$
14.5	8	$14.5 \times 8 = 116.0$
24.5	13	318.5
34.5	20	690.0
44.5	22	979.0
54.5	21	1144.5
65.0	11	715.0
Totals	**100**	**3985.5**

The mean is given by:

$$\bar{x} = \frac{\sum xf}{\sum f}$$

$$= \frac{3985.5}{100} = 39.855$$

The original raw data, summarised in the frequency table on the previous page, are shown below.

10	18	68	67	25	62	49	11	12	8
9	46	53	57	30	63	34	21	68	31
20	16	29	13	31	56	9	34	45	55
35	40	45	48	54	50	34	32	47	60
70	52	21	25	53	41	29	63	43	50
40	48	45	38	51	25	52	55	47	46
46	50	8	25	56	18	20	36	36	9
38	39	53	45	42	42	61	55	30	38
62	47	58	54	59	25	24	53	42	61
18	30	32	45	49	28	31	27	54	38

In this form it is impossible to get an overview of the number of cars, nor would listing every possible value in a frequency table (0 to 70) be helpful.

However, grouping the data and estimating the mean was not the only option. Drawing a stem-and-leaf diagram and using it to find the median would have been another possibility.

> Is it possible to find estimates for the other measures of centre?
> Find the mean of the original data and compare it to the estimate.

The data the reporter collected when researching his article on cycling accidents included the distance from home, in metres, of those involved in cycling accidents. In full these were as follows.

3000	75	1200	300	50	10	150	1500	250	25
200	4500	35	60	120	400	2400	140	45	5
1250	3500	30	75	250	1200	250	50	250	450
15	4000								

It is clear that there is considerable spread in the data. It is continuous data and the reporter is aware that they appear to have been rounded but he does not know to what level of precision. Consequently there is no way of reflecting the level of precision in setting the interval boundaries.

The reporter wants to estimate the mean and decides on the following grouping.

Location relative to home	Distance, d, in metres	Distance mid-value, x	Frequency (number of accidents), f	$x \times f$
Very close	$0 \leqslant d < 100$	50	12	600
Close	$100 \leqslant d < 500$	300	11	3 300
Not far	$500 \leqslant d < 1500$	1000	3	3 000
Quite far	$1500 \leqslant d < 5000$	3250	6	19 500
Totals			**32**	**26 400**

$$\bar{x} = \frac{26\,400}{32} = 825\,\text{m}$$

A summary of the measures of centre for the original and grouped accident data is given below.

	Raw data	Grouped data
Mean	$25\,785 \div 32 = 806\,\text{m}$	$825\,\text{m}$
Mode	$250\,\text{m}$	Modal group $0 \leqslant d < 100\,\text{m}$
Median	$\frac{1}{2}(200 + 250) = 225\,\text{m}$	

> ❯ Which measure of centre seems most appropriate for these data?

The reporter's article

The reporter decided that he had enough information and wrote the article below.

A town council that does not care

The level of civilisation of any society can be measured by how much it cares for its most vulnerable members.

On that basis our town council rates somewhere between savages and barbarians. Every day they sit back complacently while those least able to defend themselves, the very old and the very young, run the gauntlet of our treacherous streets.

I refer of course to the lack of adequate safety measures for our cyclists, 60% of whom are children or senior citizens. Statistics show that they only have to put one wheel outside their front doors to be in mortal danger. 80% of cycling accidents happen within 1500 metres of home.

Last week Rita Roy became the latest unwitting addition to these statistics. Luckily she is now on the road to recovery but that is no thanks to the members of our unfeeling town council who set people on the road to death and injury without a second thought.

What, this paper asks our councillors, are you doing about providing safe cycle tracks from our housing estates to our schools and shopping centres? And what are you doing to promote safety awareness among our cyclists, young and old?

Answer: Nothing.

⟩ Is it a fair article?

⟩ Is it justified, based on the available evidence?

Continuous data

For a statistics project Robert, a student at Avonford College, collected the heights of 50 female students.

He constructed a frequency table for his project and included the calculations to find an estimate for the mean of his data.

Height, h	Mid-value, x	Frequency, f	xf
$157 < h \leqslant 159$	158	4	632
$159 < h \leqslant 161$	160	11	1760
$161 < h \leqslant 163$	162	19	3078
$163 < h \leqslant 165$	164	8	1312
$165 < h \leqslant 167$	166	5	830
$167 < h \leqslant 169$	168	3	504
Totals		**50**	**8116**

$$\bar{x} = \frac{8116}{50}$$
$$= 162.32$$

Note: Class boundaries

His teacher was concerned about the class boundaries and asked Robert 'To what degree of accuracy have you recorded your data?' Robert told him 'I rounded all my data to the nearest centimetre'. Robert showed his teacher his raw data.

163	160	167	168	166	164	166	162	163	163
165	163	163	159	159	158	162	163	163	166
164	162	164	160	161	162	162	160	169	162
163	160	167	162	158	161	162	163	165	165
163	163	168	165	165	161	160	161	161	161

Robert's teacher said that the class boundaries should have been:

$$157.5 \leqslant h < 159.5$$
$$159.5 \leqslant h < 161.5, \text{and so on.}$$

He explained that a height recorded to the nearest centimetre as 158 cm has a value in the interval 158 ± 0.5 cm (this can be written as $157.5 \leqslant h < 158.5$). Similarly the actual values of those recorded as 159 cm lie in the interval $158.5 \leqslant h < 159.5$. So, the interval $157.5 \leqslant h < 159.5$ covers the actual values of the data items 158 and 159. The interval $159.5 \leqslant h < 161.5$ covers the actual values of 160 and 161, and so on.

> ❯ What adjustment does Robert need to make to his estimated mean in the light of his teacher's comments?
> ❯ Find the mean of the raw data. What do you notice when you compare it with your estimate?

You are not always told the level of precision of summarised data and the class widths are not always equal, as the reporter for the local newspaper discovered. Also, there are different ways of representing class boundaries, as the following example illustrates.

Example 1.4

The frequency distribution shows the lengths of telephone calls made by Emily during August. Choose suitable mid-class values and estimate Emily's mean call time for August.

Solution

Time (seconds)	Mid-value, x	Frequency, f	xf
0–	30	39	1170
60–	90	15	1350
120–	150	12	1800
180–	240	8	1920
300–	400	4	1600
500–1000	750	1	750
Totals		**79**	**8590**

$$\bar{x} = \frac{8590}{79}$$
$$= 108.7 \text{ seconds}$$

Emily's mean call time is 109 seconds, to 3 significant figures.

Notes

1 The interval '0–' can be written as $0 \leqslant x < 60$, the interval '60–' can be written as $60 \leqslant x < 120$, and so on, up to '500–1000', which can be written as $500 \leqslant x \leqslant 1000$.

2 There is no indication of the level of precision of the recorded data. They may have been recorded to the nearest second.

3 The class widths vary.

Exercise 1D

1 A college nurse keeps a record of the heights, measured to the nearest centimetre, of a group of students she treats.

Her data are summarised in the following grouped frequency table.

Height (cm)	Number of students
110–119	1
120–129	3
130–139	10
140–149	28
150–159	65
160–169	98
170–179	55
180–189	15

Choose suitable mid-class values and calculate an estimate for the mean height.

CP

2 A junior school teacher noted the times to the nearest minute a group of children spent reading during a particular day.

The data are summarised as follows.

Time (nearest minute)	Number of children
20–29	12
30–39	21
40–49	36
50–59	24
60–69	12
70–89	9
90–119	2

(i)   Choose suitable mid-class values and calculate an estimate for the mean time spent reading by the pupils.

(ii) Some time later, the teacher collected similar data from a group of 25 children from a neigbouring school. She calculated the mean to be 75.5 minutes. Compare the estimate you obtained in part (i) with this value.

What assumptions must you make for the comparison to be meaningful?

3 The stated ages of the 91 cyclists considered earlier are summarised by the following grouped frequency distribution.

Stated age (years)	Frequency
0–9	13
10–19	26
20–29	16
30–39	10
40–49	6
50–59	5
60–69	14
70–79	0
80–89	1
Total	**91**

(i)   Choose suitable mid-interval values and calculate an estimate of the mean stated age.

(ii) Make a suitable error adjustment to your answer to part (i) to give an estimate of the mean age of the cyclists.

(iii) The adjusted mean of the actual data was 30.4 years. Compare this with your answer to part (ii) and comment.

4 In an agricultural experiment, 320 plants were grown on a plot. The lengths of the stems were measured, to the nearest centimetre, 10 weeks after planting. The lengths were found to be distributed as in the following table.

Length, x (cm)	Frequency (number of plants)
$20.5 \leqslant x < 32.5$	30
$32.5 \leqslant x < 38.5$	80
$38.5 \leqslant x < 44.5$	90
$44.5 \leqslant x < 50.5$	60
$50.5 \leqslant x < 68.5$	60

Calculate an estimate of the mean of the stem lengths from this experiment.

5 The reporter for the local newspaper considered choosing different classes for the data dealing with the cyclists who were involved in accidents.

He summarised the distances from home of 32 cyclists as follows.

Distance, d (metres)	Frequency
$0 \leqslant d < 50$	7
$50 \leqslant d < 100$	5
$100 \leqslant d < 150$	2
$150 \leqslant d < 200$	1
$200 \leqslant d < 300$	5
$300 \leqslant d < 500$	3
$500 \leqslant d < 1000$	0
$1000 \leqslant d < 5000$	9
Total	**32**

(i) Choose suitable mid-class values and estimate the mean.

(ii) The mean of the raw data is 806 m and his previous grouping gave an estimate for the mean of 825 m. Compare your answer to this value and comment.

6 A crate containing 270 oranges was opened and each orange was weighed. The masses, given to the nearest gram, were grouped and the resulting distribution is as follows.

Mass, x (grams)	Frequency (number of oranges)
60–99	20
100–119	60
120–139	80
140–159	50
160–220	60

(i) State the class boundaries for the interval 60–99.

(ii) Calculate an estimate for the mean mass of the oranges from the crate.

1.8 Measures of spread (variation)

In the last section you saw how an estimate for the mean can be found from grouped data. The mean is just one example of a **typical value** of a data set. You also saw how the mode and the median can be found from small data sets. The next chapter considers the use of the median as a typical value when dealing with grouped data and also the **interquartile range** as a **measure of spread**. In this chapter we will consider the range, the mean absolute deviation, the variance and the standard deviation as measures of spread.

Range

The simplest measure of spread is the **range**. This is just the difference between the largest value in the data set (the upper extreme) and the smallest value (the lower extreme).

» Range = largest − smallest

The figures below are the prices, in cents, of a 100 g jar of *Nesko* coffee in ten different shops.

161 161 163 163 167 168 170 172 172 172

The range for this data is:

Range = 172 − 161 = 11 cents

Example 1.5

Ruth is investigating the amount of money, in dollars, students at Avonford College earn from part-time work on one particular weekend. She collects and orders data from two classes and this is shown below.

Class 1	Class 2
10 10 10 10 10 10 12 15 15 15	10 10 10 10 10 10 12 12 12 12
16 16 16 16 18 18 20 25 38 90	15 15 15 15 16 17 18 19 20 20
	25 35 35

She calculates the mean amount earned for each class. Her results are:

Class 1: $\bar{x}_1 = \$19.50$

Class 2: $\bar{x}_2 = \$16.22$

She concludes that the students in Class 1 each earn about $3 more, on average, than the students in Class 2.

Her teacher suggests she look at the spread of the data. What further information does this reveal?

Solution

Ruth calculates the range for each class: Range (Class 1) = $80
Range (Class 2) = $25

She concludes that the part-time earnings in Class 1 are much more spread out.

However, when Ruth looks again at the raw data she notices that one student in Class 1 earned $90, considerably more than anybody else. If that item of data is ignored then the spread of data for the two classes is similar.

One of the problems with the range is that it is prone to the effect of extreme values.

> Calculate the mean earnings of Class 1 with the item $90 removed.

> What can you conclude about the effect of extreme values on the mean?

The range does not use all of the available information; only the extreme values are used. In quality control this can be an advantage as it is very sensitive to something going wrong on a production line. Also the range is easy to calculate. However, usually we want a measure of spread that uses all the available data and that relates to a central value.

The mean absolute deviation

Kim and Joe play as strikers for two local football teams. They are being considered for the state team. The team manager is considering their scoring records.

Kim's scoring record over ten matches looks like this:

$$0 \quad 1 \quad 0 \quad 3 \quad 0 \quad 2 \quad 0 \quad 0 \quad 0 \quad 4$$

Joe's record looks like this:

$$1 \quad 1 \quad 1 \quad 0 \quad 0 \quad 2 \quad 1 \quad 1 \quad 2 \quad 2$$

The mean scores are, for Kim, $\bar{x}_1 = 1$ and, for Joe, $\bar{x}_2 = 1.1$.

Looking first at Kim's data consider the differences, or *deviations*, of his scores from the mean.

Number of goals scored, x	0	1	0	3	0	2	0	0	0	4
Deviations $(x - \bar{x})$	−1	0	−1	2	−1	1	−1	−1	−1	3

To find a summary measure you need to combine the deviations in some way. If you just add them together they total zero.

> Why does the sum of the deviations always total zero?

The mean absolute deviation ignores the signs and adds together the **absolute deviations**. The symbol $|d|$ tells you to take the positive, or absolute, value of d.

For example $|-2| = 2$ and $|2| = 2$.

It is now possible to sum the deviations:

$$1 + 0 + 1 + 2 + 1 + 1 + 1 + 1 + 1 + 3 = 12,$$

the **total of the absolute deviations**.

It is important that any measure of spread is not linked to the sample size so you have to average out this total by dividing by the sample size.

In this case the sample size is 10. The **mean absolute deviation** $= \dfrac{12}{10} = 1.2$.

» The mean absolute deviation from the mean $= \dfrac{1}{n}\sum |x - \bar{x}|$

> Remember
> $n = \Sigma f$.

For Joe's data the mean absolute deviation is:

$$\tfrac{1}{10}(0.1 + 0.1 + 0.1 + 1.1 + 1.1 + 0.9 + 0.1 + 0.1 + 0.9 + 0.9) = 0.54$$

The average numbers of goals scored by Kim and Joe are similar (1.0 and 1.1) but Joe is less variable (or more consistent) in his goal scoring (0.54 compared to 1.2).

The mean absolute deviation is an acceptable measure of spread but is not widely used because it is difficult to work with. The **standard deviation** is more important mathematically and is more extensively used.

The variance and standard deviation

An alternative to ignoring the signs is to square the differences or deviations. This gives rise to a measure of spread called the **variance**, which when square-rooted gives the standard deviation.

Though not as easy to calculate as the mean absolute deviation, the standard deviation has an important role in the study of more advanced statistics.

To find the variance of a data set: For Kim's data this is:

» Square the deviations $(x - \bar{x})^2$ $(0 - 1)^2, (1 - 1)^2, (0 - 1)^2$, etc.

» Sum the squared deviations $\sum(x - \bar{x})^2$ $1 + 0 + 1 + 4 + 1 + 1 +$
$1 + 1 + 1 + 9 = 20$

» Find their mean $\dfrac{\sum(x - \bar{x})^2}{n}$ $\dfrac{20}{10} = 2$

This is known as the variance.

» Variance $= \dfrac{\sum(x - \bar{x})}{n}$

The square root of the variance is called the standard deviation.

» $sd = \sqrt{\dfrac{\sum(x - \bar{x})^2}{n}}$

> So, for Kim's data the variance is 2, but what are the units? In calculating the variance the data are squared. In order to get a measure of spread that has the same units as the original data it is necessary to take the square root of the variance. The resulting statistical measure is known as the *standard deviation*.

In other books or on the internet, you may see this calculation carried out using $n-1$ rather than n as the divisor. In this case the answer is denoted by s.

$$s = \sqrt{\frac{\sum(x-\bar{x})^2}{n-1}}$$

In *Probability & Statistics 1*, you should always use n as the divisor. You will meet s if you go on to study *Probability & Statistics 2*.

So for Kim's data the variance is 2, *sd* is $\sqrt{2} = 1.41$ (to 3 s.f.).

This example, using Joe's data, shows how the variance and standard deviation are calculated when the data are given in a frequency table. We've already calculated the mean; $\bar{x} = 1.1$.

Number of goals scored, x	Frequency, f	Deviation $(x - \bar{x})$	Deviation^2 $(x - \bar{x})^2$	Deviation$^2 \times f$ $[(x - \bar{x})^2 f]$
0	2	$0 - 1.1 = -1.1$	1.21	$1.21 \times 2 = 2.42$
1	5	$1 - 1.1 = -0.1$	0.01	$0.01 \times 5 = 0.05$
2	3	$2 - 1.1 = 0.9$	0.81	$0.81 \times 3 = 2.43$
Totals	**10**			**4.90**

For data presented in this way,

$$\text{standard deviation} = \sqrt{\frac{\sum(x-\bar{x})^2 f}{n}} = \sqrt{\frac{\sum(x-\bar{x})^2 f}{\sum f}} \qquad \boxed{\Sigma f = n}$$

The standard deviation for Joe's data is $sd = \sqrt{\dfrac{4.90}{10}} = 0.7$ goals.

Comparing this to the standard deviation of Kim's data (1.41), we see that Joe's goal scoring is more consistent (or less variable) than Kim's. This confirms what was found when the mean absolute deviation was calculated for each data set. Joe was found to be a more consistent scorer (mean absolute deviation = 0.54) than Kim (mean absolute deviation = 1.2).

An alternative form for the standard deviation

The arithmetic involved in calculating $\sum(x-\bar{x})^2 f$ can often be very messy.

An alternative formula for calculating the standard deviation is given by:

» standard deviation $= \sqrt{\dfrac{\sum x^2 f}{n} - \bar{x}^2}$ or $\sqrt{\dfrac{\sum x^2 f}{\sum f} - \bar{x}^2}$

Consider Joe's data one more time.

Number of goals scored, x	Frequency, f	xf	x^2f
0	2	0	0
1	5	5	5
2	3	6	12
Total	**10**	**11**	**17**

$$\bar{x} = \frac{11}{10}$$

$$= 1.1$$

$$\text{standard deviation} = \sqrt{\frac{17}{10} - 1.1^2}$$

$$= \sqrt{1.7 - 1.21}$$

$$= \sqrt{0.49}$$

$$= 0.7$$

This gives the same result as using $\sqrt{\dfrac{\sum (x - \bar{x})^2 f}{\sum f}}$. The derivation of this alternative form for the standard deviation is given in Appendix 1 at www.hoddereducation.com/cambridgeextras.

> In practice you will make extensive use of your calculator's statistical functions to find the mean and standard deviation of sets of data.
>
> Care should be taken as the notations S, s, sd, σ and $\hat{\sigma}$ are used differently by different calculator manufacturers, authors and users. You will meet σ in Chapter 4.

The following examples involve finding or using the sample variance.

Example 1.6

Find the mean and the standard deviation of a sample with

$$\sum x = 960, \sum x^2 = 18\,000, n = 60.$$

Solution

$$\bar{x} = \frac{\sum x}{n} = \frac{960}{60} = 16$$

$$\text{variance} = \frac{\sum x^2}{n} - \bar{x}^2 = \frac{18\,000}{60} - 16^2 = 44$$

$$\text{standard deviation} = \sqrt{44} = 6.63 \text{ (to 3 s.f.)}$$

Example 1.7

Find the mean and the standard deviation of a sample with

$$\sum(x-\bar{x})^2 = 2000, \sum x = 960, \sum f = 60.$$

Solution

$$\bar{x} = \frac{960}{60} = 16$$

Remember:
$\Sigma f = n$

$$\text{variance} = \frac{\sum(x-\bar{x})^2}{\sum f} = \frac{2000}{60} = 33.3\ldots$$

$$\text{standard deviation} = \sqrt{33.3\ldots} = 5.77 \text{ (to 3 s.f.)}$$

Example 1.8

As part of her job as quality controller, Stella collected data relating to the life expectancy of a sample of 60 light bulbs produced by her company. The mean life was 650 hours and the standard deviation was 8 hours. A second sample of 80 bulbs was taken by Sol and resulted in a mean life of 660 hours and standard deviation 7 hours.

Find the overall mean and standard deviation.

Solution

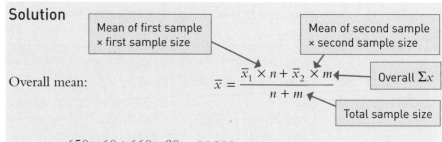

Mean of first sample
× first sample size

Mean of second sample
× second sample size

Overall mean:

$$\bar{x} = \frac{\bar{x}_1 \times n + \bar{x}_2 \times m}{n + m}$$

Overall Σx

Total sample size

$$\bar{x} = \frac{650 \times 60 + 660 \times 80}{60 + 80} = \frac{91\,800}{140} = 655.71\ldots = 656 \text{ hours (to 3 s.f.)}$$

For Stella's sample the variance is 8^2. Therefore $8^2 = \dfrac{\sum x_1^2}{60} - 650^2$.

For Sol's sample the variance is 7^2. Therefore $7^2 = \dfrac{\sum x_2^2}{80} - 660^2$.

From the above Stella found that

$$\sum x_1^2 = (8^2 + 650^2) \times 60 = 25\,353\,840 \text{ and } \sum x_2^2 = 34\,851\,920.$$

The overall variance is

Overall Σx^2

Do not round any numbers until you have completed all calculations.

The total number of light bulbs is 140.

$$\frac{25\,353\,840 + 34\,851\,920}{140} - 655.71\ldots^2$$

$$= 430\,041.14\ldots - 429\,961.22\ldots$$

$$= 79.91\ldots$$

The overall standard deviation is $\sqrt{79.91\ldots} = 8.94$ hours (to 3 s.f.).

> ❯ Carry out the calculation in Example 1.8 using rounded numbers. That is, use 656 for the overall mean rather than 655.71….What do you notice?

The standard deviation and outliers

Data sets may contain extreme values and when this occurs you are faced with the problem of how to deal with them.

Many data sets are samples drawn from parent populations which are normally distributed. You will learn more about the normal distribution in Chapter 7. In these cases approximately:

➤➤ 68% of the values lie within 1 standard deviation of the mean

➤➤ 95% lie within 2 standard deviations of the mean

➤➤ 99.75% lie within 3 standard deviations of the mean.

If a particular value is *more than two standard deviations from the mean* it should be investigated as possibly not belonging to the data set. If it is as much as three standard deviations or more from the mean then the case to investigate it is even stronger.

> The 2-standard-deviation test should not be seen as a way of defining outliers. It is only a way of identifying those values which it might be worth looking at more closely.

In an A Level Spanish class the examination marks at the end of the year are shown below.

| 35 | 52 | 55 | 61 | 96 | 63 | 50 | 58 | 58 | 49 | 61 |

The value 96 was thought to be significantly greater than the other values. The mean and standard deviation of the data are $\bar{x} = 58$ and $sd = 14.16...$. The value 96 is more than two standard deviations above the mean:

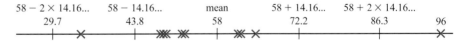

▲ Figure 1.9

Further investigation revealed that the mark of 96 was achieved by a Spanish boy who had taken A Level Spanish because he wanted to study Spanish at university. It might be appropriate to omit this value from the data set.

> ❯ Calculate the mean and standard deviation of the data with the value 96 left out.
> ❯ Investigate the value using your new mean and standard deviation.

The times taken, in minutes, for some train journeys between Kolkata and Majilpur were recorded as shown.

56	61	57	55	58	57	5	60	61	59

It is unnecessary here to calculate the mean and standard deviation. The value 5 minutes is obviously a mistake and should be omitted unless it is possible to correct it by referring to the original source of data.

Exercise 1E

1 (i)   Find the mean of the following data.

0 0 0 1 1 1 1 1 2 2 2 2 2 2 3 3 3 3 4 4 4 4 5 5

(ii)   Find the standard deviation using both forms of the formula.

2   Find the mean and standard deviation of the following data.

x	3	4	5	6	7	8	9
f	2	5	8	14	9	4	3

CP 3   Mahmood and Raheem are football players. In the 30 games played so far this season their scoring records are as follows.

Goals scored	0	1	2	3	4
Frequency (Mahmood)	12	8	8	1	1
Frequency (Raheem)	4	21	5	0	0

(i)   Find the mean and the standard deviation of the number of goals each player scored.

(ii)   Comment on the players' goal scoring records.

4   For a set of 20 items of data $\sum x = 22$ and $\sum x^2 = 55$. Find the mean and the standard deviation of the data.

5   For a data set of 50 items of data $\sum (x - \bar{x})^2 f = 8$ and $\sum xf = 20$. Find the mean and the standard deviation of the data.

CP 6   Two thermostats were used under identical conditions. The water temperatures, in °C, are given below.

Thermostat A: 24   25   27   23   26

Thermostat B: 26   26   23   22   28

(i)   Calculate the mean and standard deviation for each set of water temperatures.

(ii)   Which is the better thermostat? Give a reason.

A second sample of data was collected using thermostat A.

25   24   24   25   26   25   24   24

(iii)   Find the overall mean and the overall standard deviation for the two sets of data for thermostat A.

CP **7** Ditshele has a choice of routes to work. She timed her journey along each route on several occasions and the times in minutes are given below.

Town route: 15   16   20   28   21

Country route: 19   21   20   22   18

(i)   Calculate the mean and standard deviation of each set of journey times.

(ii) Which route would you recommend? Give a reason.

M **8** In a certain district, the mean annual rainfall is 80 cm, with standard deviation 4 cm.

(i)   One year it was 90 cm. Was this an exceptional year?

(ii) The next year had a total of 78 cm. Was that exceptional?

Jake, a local amateur meteorologist, kept a record of the weekly rainfall in his garden. His first data set, comprising 20 weeks of figures, resulted in a mean weekly rainfall of 1.5 cm. The standard deviation was 0.1 cm. His second set of data, over 32 weeks, resulted in a mean of 1.7 cm and a standard deviation of 0.09 cm.

(iii) Calculate the overall mean and the overall standard deviation for the whole year.

(iv) Estimate the annual rainfall in Jake's garden.

M **9** A farmer expects to harvest a crop of 3.8 tonnes, on average, from each hectare of his land, with standard deviation 0.2 tonnes.

One year there was much more rain than usual and he harvested 4.1 tonnes per hectare.

(i)   Was this exceptional?

(ii) Do you think the crop was affected by the unusual weather or was the higher yield part of the variability which always occurs?

M **10** A machine is supposed to produce ball bearings with a mean diameter of 2.0 mm. A sample of eight ball bearings was taken from the production line and the diameters measured. The results, in millimetres, were as follows:

2.0   2.1   2.0   1.8   2.4   2.3   1.9   2.1

(i)   Calculate the mean and standard deviation of the diameters.

(ii) Do you think the machine is correctly set?

PS 11 On page 28 you saw the example about Robert, the student at Avonford College who collected data relating to the heights of female students. This is his corrected frequency table and his calculations so far.

Height, h	Mid–value, x	Frequency, f	xf
$157.5 \leqslant h < 159.5$	158.5	4	634.0
$159.5 \leqslant h < 161.5$	160.5	11	1765.5
$161.5 \leqslant h < 163.5$	162.5	19	3087.5
$163.5 \leqslant h < 165.5$	164.5	8	1316.0
$165.5 \leqslant h < 167.5$	166.5	5	832.5
$167.5 \leqslant h < 169.5$	168.5	3	505.5
Totals		**50**	**8141.0**

$$\bar{x} = \frac{8141.0}{50} = 162.82$$

(i) Calculate the standard deviation.

Robert's friend Asha collected a sample of heights from 50 male PE students. She calculated the mean and standard deviation to be 170.4 cm and 2.50 cm. Later on they realised they had excluded two measurements. It was not clear to which of the two data sets, Robert's or Asha's, the two items of data belonged. The values were 171 cm and 166 cm. Robert felt confident about one of the values but not the other.

(ii) Investigate and comment.

PS 12 As part of a biology experiment Thabo caught and weighed 120 minnows. He used his calculator to find the mean and standard deviation of their weights.

Mean   26.231 g

Standard deviation   4.023 g

(i) Find the total weight, $\sum x$, of Thabo's 120 minnows.

(ii) Use the formula standard deviation $= \sqrt{\dfrac{\sum x^2}{n} - \bar{x}^2}$ to find $\sum x^2$ for Thabo's minnows.

Another member of the class, Sharon, did the same experiment with minnows caught from a different stream. Her results are summarised by:

$$n = 80 \quad \bar{x} = 25.214 \quad \text{standard deviation} = 3.841$$

Their teacher says they should combine their results into a single set but they have both thrown away their measurements.

(iii) Find n, $\sum x$ and $\sum x^2$ for the combined data set.

(iv) Find the mean and standard deviation for the combined data set.

13 The table shows the mean and standard deviation of the weights of some turkeys and geese.

	Number of birds	Mean (kg)	Standard deviation (kg)
Turkeys	9	7.1	1.45
Geese	18	5.2	0.96

(i) Find the mean weight of the 27 birds.

(ii) The weights of individual turkeys are denoted by x_t kg and the weights of individual geese by x_g kg. By first finding $\sum x_t^2$ and $\sum x_g^2$, find the standard deviation of the weights of all 27 birds.

Cambridge International AS & A Level Mathematics
9709 Paper 61 Q5 June 2015

14 A group of 10 women and 13 men found that the mean age $\bar{x}_w$ of the 10 women was 41.2 years and the standard deviation of the women's ages was 15.1 years. For the 13 men, the mean age $\bar{x}_m$ was 46.3 years and the standard deviation was 12.7 years.

(i) Find the mean age of the whole group of 23 people.

(ii) The individual women's ages are denoted by x_w and the individual men's ages by x_m. By first finding $\sum x_w^2$ and $\sum x_m^2$, find the standard deviation for the whole group.

Cambridge International AS & A Level Mathematics
9709 Paper 6 Q4 November 2005

15 The numbers of rides taken by two students, Fei and Graeme, at a fairground are shown in the following table.

Student	Roller coaster	Water slide	Revolving drum
Fei	4	2	0
Graeme	1	3	6

(i) The mean cost of Fei's rides is $2.50 and the standard deviation of the costs of Fei's rides is $0. Explain how you can tell that the roller coaster and the water slide each cost $2.50 per ride.

(ii) The mean cost of Graeme's rides is $3.76. Find the standard deviation of the costs of Graeme's rides.

Cambridge International AS & A Level Mathematics
9709 Paper 61 Q4 June 2010

1.9 Working with an assumed mean

Human computer has it figured   Mathman Web

Schoolboy Simon Newton astounded his classmates and their parents at a school open evening when he calculated the average of a set of numbers in seconds while everyone else struggled with their adding up.

Mr Truscott, a parent of one of the other children, said, 'I was still looking for my calculator when Simon wrote the answer on the board'.

Simon modestly said when asked about his skill 'It's simply a matter of choosing a good assumed mean'.

Mathman Web wants to know 'What is the secret method, Simon?'

Without a calculator, see if you can match Simon's performance. The data is repeated below. Send your result and how you did it into Mathman Web. Don't forget – no calculators!

Number	Frequency
3510	6
3512	4
3514	3
3516	1
3518	2
3520	4

Simon gave a big clue about how he calculated the mean so quickly. He said 'It's simply a matter of choosing a good assumed mean'. Simon noticed that subtracting 3510 from each value simplified the data significantly. This is how he did his calculations.

Number, x	Number – 3510, y	Frequency, f	$y \times f$
3510	0	6	$0 \times 6 = 0$
3512	2	4	$2 \times 4 = 8$
3514	4	3	$4 \times 3 = 12$
3516	6	1	$6 \times 1 = 6$
3518	8	2	$8 \times 2 = 16$
3520	10	4	$10 \times 4 = 40$
Totals		**20**	**82**

Average (mean) $= \dfrac{82}{20} = 4.1$

(3510 is now added back) $3510 + 4.1 = 3514.1$

Simon was using an **assumed mean** to ease his arithmetic.

Sometimes it is easier to work with an assumed mean in order to find the standard deviation.

Example 1.9

Using an assumed mean of 7, find the true mean and the standard deviation of the data set 5, 7, 9, 4, 3, 8.

> It doesn't matter if the assumed mean is not very close to the correct value for the mean, but the closer it is the simpler the working will be.

Solution

Let d represent the variation from the assumed mean of 7. So $d = x - 7$.

x	$d = x - 7$	$d^2 = (x - 7)^2$
5	$5 - 7 = -2$	4
7	$7 - 7 = 0$	0
9	$9 - 7 = 2$	4
4	$4 - 7 = -3$	9
3	$3 - 7 = -4$	16
8	$8 - 7 = 1$	1
Totals	$\sum d = \sum(x - 7) = -6$	$\sum d^2 = \sum(x - 7)^2 = 34$

The mean of d is given by

$$\bar{d} = \frac{\sum d}{n} = \frac{-6}{6} = -1$$

The standard deviation of d is given by

$$sd_d = \sqrt{\frac{\sum d^2}{n} - \left(\frac{\sum d}{n}\right)^2} = \sqrt{\frac{34}{6} - \left(\frac{-6}{6}\right)^2} \longleftarrow \boxed{\bar{d}^2}$$

$$= 2.16 \text{ to 3 s.f.}$$

So the true mean is $7 - 1 = 6$.

The true standard deviation is 2.16 to 3 s.f.

> 7 is the assumed mean.

In general:

» $\bar{x} = a + \bar{d}$ where a is the assumed mean.
» the standard deviation of x is the standard deviation of d.

The following example uses summary statistics, rather than the raw data values.

Example 1.10

For a set of 10 data items, $\sum(x-9) = 7$ and $\sum(x-9)^2 = 17$.

Find their mean and standard deviation.

Solution

Let $x - 9 = d$

$$\sum(x-9) = 7 \qquad \sum d = 7$$

$$\bar{d} = \frac{7}{10} = 0.7$$

$$\bar{x} = 9 + 0.7 = 9.7$$

The mean of x is 9.7.

The assumed mean is 9.

The standard deviation of $d = \sqrt{\dfrac{\sum d^2}{n} - \left(\dfrac{\sum d}{n}\right)^2}$ where $d = x - 9$

$$= \sqrt{\frac{17}{10} - \left(\frac{7}{10}\right)^2}$$

$$= \sqrt{1.7 - 0.49}$$

$$= \sqrt{1.21}$$

$$= 1.1$$

Since the standard deviation of x is equal to the standard deviation of d, it follows that the standard deviation of x is 1.1.

The next example shows you how to use an assumed mean with grouped data.

Example 1.11

Using 162.5 as an assumed mean, find the mean and standard deviation of the data in this table. (These are Robert's figures for the heights of female students.)

Height, x (cm) midpoints	Frequency, f
158.5	4
160.5	11
162.5	19
164.5	8
166.5	5
168.5	3
Total	**50**

Solution

The working is summarised in the following table.

Height, x (cm) midpoints	$d = x - 162.5$	Frequency, f	df	d^2f
158.5	−4	4	−16	64
160.5	−2	11	−22	44
162.5	0	19	0	0
164.5	2	8	16	32
166.5	4	5	20	80
168.5	6	3	18	108
Totals		50	16	328

$$\bar{d} = \frac{16}{50} = 0.32$$

$$(sd_d)^2 = \frac{328}{50} - 0.32^2 = 6.4576$$

$$sd_d = 2.54 \text{ to 3 s.f.}$$

So the mean and standard deviation of the original data are:

$$\bar{x} = 162.5 + 0.32 = 162.82$$

$$sd_x = 2.54 \text{ to 3 s.f.}$$

Exercise 1F

1 Calculate the mean and standard deviation of the following masses, measured to the nearest gram, using a suitable assumed mean.

Mass (g)	241–244	245–248	249–252	253–256	257–260	261–264
Frequency	4	7	14	15	7	3

2 A production line produces steel bolts which have a nominal length of 95 mm. A sample of 50 bolts is taken and measured to the nearest 0.1 mm. Their deviations from 95 mm are recorded in tenths of a millimetre and summarised as $\sum x = -85$, $\sum x^2 = 734$. (For example, a bolt of length 94.2 mm would be recorded as −8.)

(i) Find the mean and standard deviation of the x values.

(ii) Find the mean and standard deviation of the lengths of the bolts in millimetres. Give the mean to 4 significant figures.

(iii) One of the figures recorded is −18. Suggest why this can be regarded as an outlier.

(iv) The figure of −18 is thought to be a mistake in the recording. Calculate the new mean (to 4 significant figures) and standard deviation of the lengths in millimetres, with the −18 value removed.

3 A system is used at a college to predict a student's A Level grade in a particular subject using their GCSE results. The GCSE score is g and the predicted A Level score is a and for Maths in 2018 the equation of the line of best fit relating them was $a = 2.6g - 9.42$.

This year there are 66 second-year students and their GCSE scores are summarised as $\Sigma g = 408.6$, $\Sigma g^2 = 2545.06$.

(i) Find the mean and standard deviation of the GCSE scores.

(ii) Find the mean of the predicted A Level scores using the 2018 line of best fit.

4 (i) Find the mode, mean and median of:

2 8 6 5 4 5 6 3 6 4 9 1 5 6 5

Hence write down, without further working, the mode, mean and median of:

(ii) 20 80 60 50 40   50 60 30 60 40   90 10 50 60 50

(iii) 12 18 16 15 14   15 16 13 16 14   19 11 15 16 15

(iv) 4 16 12 10 8   10 12 6 12 8   18 2 10 12 10

CP **5** A manufacturer produces electrical cable which is sold on reels. The reels are supposed to hold 100 metres of cable. In the quality control department the length of cable on randomly chosen reels is measured. These measurements are recorded as deviations, in centimetres, from 100 m. (So, for example, a length of 99.84 m is recorded as −16.)

For a sample of 20 reels the recorded values, x, are summarised by:

$$\sum x = -86 \qquad \sum x^2 = 4281$$

(i) Calculate the mean and standard deviation of the values of x.

(ii) Later it is noticed that one of the values of x is −47, and it is thought that so large a value is likely to be an error. Give a reason to support this view.

(iii) Find the new mean and standard deviation of the values of x when the value −47 is discarded.

6 On her summer holiday, Felicity recorded the temperatures at noon each day for use in a statistics project. The values recorded, f degrees Fahrenheit, were as follows, correct to the nearest degree.

47 59 68 62 49 67 66 73 70 68 74 84 80 72

(i) Represent Felicity's data on a stem-and-leaf diagram. Comment on the shape of the distribution.

(ii) Using a suitable assumed mean, find the mean and standard deviation of Felicity's data.

7 For a set of ten data items, $\sum(x - 20) = -140$ and $\sum(x - 20)^2 = 2050$. Find their mean and standard deviation.

8 For a set of 20 data items, $\sum(x+3) = 140$ and $\sum(x+3)^2 = 1796$. Find their mean and standard deviation.

9 For a set of 15 data items, $\sum(x+a) = 156$ and $\sum(x+a)^2 = 1854$. The mean of these values is 5.4.

Find the value of a and the standard deviation.

10 For a set of 10 data items, $\sum(x-a) = -11$ and $\sum(x-a)^2 = 75$. The mean of these values is 5.9.

Find the value of a and the standard deviation.

11 The length of time, t minutes, taken to do the crossword in a certain newspaper was observed on 12 occasions. The results are summarised below.

$$\sum(t-35) = -15 \quad \sum(t-35)^2 = 82.23$$

Calculate the mean and standard deviation of these times taken to do the crossword.

Cambridge International AS & A Level Mathematics
9709 Paper & Q1 June 2007

12 A summary of 24 observations of x gave the following information:

$$\sum(x-a) = -73.2 \quad \text{and} \quad \sum(x-a)^2 = 2115.$$

The mean of these values of x is 8.95.

(i) Find the value of the constant a.

(ii) Find the standard deviation of these values of x.

Cambridge International AS & A Level Mathematics
9709 Paper 6 Q1 November 2007

KEY POINTS

1 An item of data x may be identifed as an *outlier* if

$$|x - \bar{x}| > 2 \times \text{standard deviation.}$$

That is, if x is more than two standard deviations above or below the sample mean.

2 *Categorical data* are non-numerical; *discrete data* can be listed; *continuous data* can be measured to any degree of accuracy and it is not possbile to list all values.

3 *Stem-and-leaf diagrams* (or stemplots) are suitable for discrete or continuous data. All data values are retained as well as indicating properties of the distribution.

4 The mean, median and the mode or modal class are measures of central tendency.

5 *The mean,* $\bar{x} = \dfrac{\sum x}{n}$. For grouped data $\bar{x} = \dfrac{\sum xf}{\sum f}$.

6 The *median* is the mid-value when the data are presented in rank order; it is the value of the $\dfrac{n+1}{2}$th item of n data items.

7 The *mode* is the most common item of data. The *modal* class is the class containing the most data, when the classes are of equal width.

8 The range, variance and standard deviation are measures of *spread* or *variation* or *dispersion*.

9 *Range* = maximum data value − minimum data value.

10 The standard deviation = $\sqrt{\dfrac{\sum (x - \bar{x})^2 f}{n}}$ or $\sqrt{\dfrac{\sum (x - \bar{x})^2 f}{\sum f}}$

11 An alternative form is

standard deviation = $\sqrt{\dfrac{\sum x^2 f}{n} - \bar{x}^2}$ or $\sqrt{\dfrac{\sum x^2 f}{\sum f} - \bar{x}^2}$

12 Working with an assumed mean a,

$$\bar{x} = a + \bar{d}$$

where a is the assumed mean and d is the deviation from the assumed mean and the standard deviation of x is the standard deviation of d.

LEARNING OUTCOMES

Now that you have finished this chapter, you should be able to
- draw and interpret stem-and-leaf diagrams and back-to-back stem-and-leaf diagrams
- understand the terms:
  - variable
  - discrete data
  - continuous data
  - frequency distribution
- calculate and use measures of central tendency:
  - mean
  - median
  - mode

- calculate and use measures of variation:
  - range
  - variance
  - standard deviation
- work with raw data
- work with grouped data
- work with raw data and grouped data
- use summary statistics (Σx and Σx^2) to calculate the mean and standard deviation
- work with an assumed mean to find the mean and standard deviation from coded totals (coded totals $\Sigma(x - a)$ and $\Sigma(x - a)^2$)
- find the mean and standard deviation of data drawn from two samples.

2

Representing and interpreting data

A picture is worth a thousand numbers.
Anon

Latest news from Alpha High

The Psychology department at Alpha High have found that girls have more online friends than boys. Mr Rama, the head of department, said 'The results are quite marked, girls in all age groups with the exception of the youngest age group had significantly more online friends. We have several hypotheses to explain this, but want to do more research before we draw any conclusions.' The group of student psychologists have won first prize in a competition run by *Psychology Now* for their research.

The Psychology department are now intending to compare these results with the number of friends that students have 'real-life' contact with.

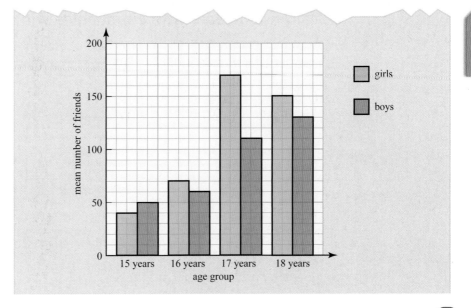

> ❯ What is the mean number of friends for 17-year-old girls?
> ❯ 220 students aged 17 were surveyed from Alpha High. 120 of these students were girls. What is the overall mean number of friends for the 17-year-olds?

Most raw data need to be **summarised** to make it easier to see any patterns that may be present. You will often want to draw a diagram too. The Psychology department used the following table to construct the diagram for their article.

Age group	15 years	16 years	17 years	18 years
Sample size	200	170	220	310
Mean number of friends – girls	40	70	170	150
Mean number of friends – boys	50	60	110	130

You will often want to use a diagram to communicate statistical findings. People find diagrams are a very useful and easy way of presenting and understanding statistical information.

2.1 Histograms

Histograms are used to illustrate continuous data. The columns in a histogram may have different widths and the area of each column is proportional to the frequency. Unlike bar charts, there are no gaps between the columns because where one class ends, the next begins.

Continuous data with equal class widths

A sample of 60 components is taken from a production line and their diameters, d mm, recorded. The resulting data are summarised in the following frequency table.

Diameter (mm)	Frequency
$25 \leqslant d < 30$	1
$30 \leqslant d < 35$	3
$35 \leqslant d < 40$	7
$40 \leqslant d < 45$	15
$45 \leqslant d < 50$	17
$50 \leqslant d < 55$	10
$55 \leqslant d < 60$	5
$60 \leqslant d < 65$	2

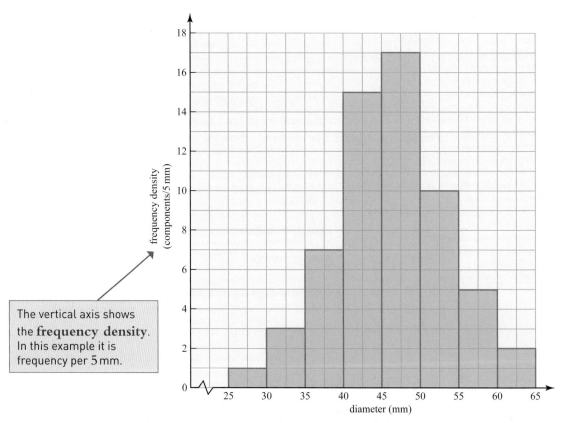

The vertical axis shows the **frequency density**. In this example it is frequency per 5 mm.

▲ **Figure 2.1** Histogram to show the distribution of component diameters

The class boundaries are 25, 30, 35, 40, 45, 50, 55, 60 and 65. The width of each class is 5.

The area of each column is proportional to the class frequency. In this example the class widths are equal so the height of each column is also proportional to the class frequency.

The column representing $45 \leqslant d < 50$ is the highest and this tells you that this is the modal class, that is the class with highest frequency per 5 mm.

> › How would you identify the modal class if the intervals were not of equal width?

Labelling the frequency axis

The vertical axis tells you the **frequency density**. Figure 2.3 looks the same as Figure 2.2 but it is not a histogram. This type of diagram is, however, often incorrectly referred to as a histogram. It is more correctly called a frequency chart. A histogram shows the frequency density on the vertical axis.

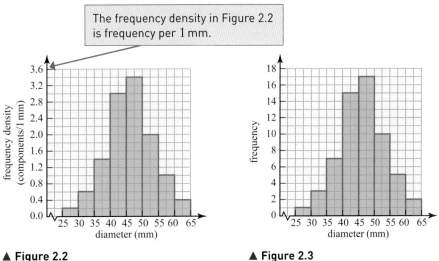

The frequency density in Figure 2.2 is frequency per 1 mm.

▲ **Figure 2.2**   ▲ **Figure 2.3**

Comparing Figures 2.2 and 2.3, you will see that the shape of the distribution remains the same but the values on the vertical axes are different. This is because different units have been used for the frequency density.

Continuous data with unequal class widths

The heights of 80 broad bean plants were measured, correct to the nearest centimetre, ten weeks after planting. The data are summarised in the following frequency table.

Height (cm)	Frequency	Class width (cm)	Frequency density
$7.5 \leqslant x < 11.5$	1	4	0.25
$11.5 \leqslant x < 13.5$	3	2	1.5
$13.5 \leqslant x < 15.5$	7	2	3.5
$15.5 \leqslant x < 17.5$	11	2	5.5
$17.5 \leqslant x < 19.5$	19	2	9.5
$19.5 \leqslant x < 21.5$	14	2	7
$21.5 \leqslant x < 23.5$	13	2	6.5
$23.5 \leqslant x < 25.5$	9	2	4.5
$25.5 \leqslant x < 28.5$	3	3	1

When the class widths are unequal you can use

$$\text{frequency density} = \frac{\text{frequency}}{\text{class width}}$$

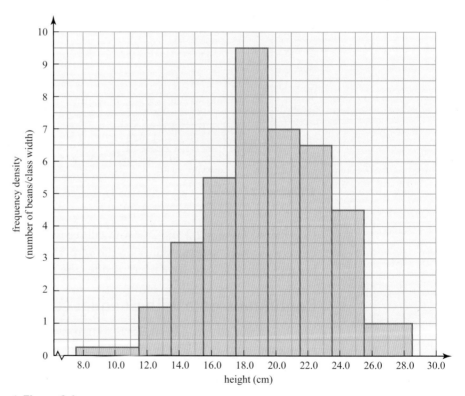

▲ Figure 2.4

Discrete data

> Histograms are occasionally used for grouped **discrete** data. However, you should always first consider the alternatives. ❗

A test was given to 100 students. The maximum mark was 70. The raw data are shown below.

10	18	68	67	25	62	49	11	12	8
9	46	53	57	30	63	34	21	68	31
20	16	29	13	31	56	9	34	45	55
35	40	45	48	54	50	34	32	47	60
70	52	21	25	53	41	29	63	43	50
40	48	45	38	51	25	52	55	47	46
46	50	8	25	56	18	20	36	36	9
38	39	53	45	42	42	61	55	30	38
62	47	58	54	59	25	24	53	42	61
18	30	32	45	49	28	31	27	54	38

Illustrating this data using a vertical line graph results in Figure 2.5.

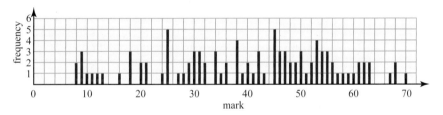

▲ Figure 2.5

This diagram fails to give a clear picture of the overall distribution of marks. In this case you could consider a bar chart or, as the individual marks are known, a stem-and-leaf diagram, as follows.

$n = 100$

2 | 5 represents 25 marks

```
0 | 8 8 9 9 9
1 | 0 1 2 3 6 8 8 8
2 | 0 0 1 1 4 5 5 5 5 5 7 8 9 9
3 | 0 0 0 1 1 1 2 2 4 4 4 5 6 6 8 8 8 8 9
4 | 0 0 1 2 2 2 3 5 5 5 5 5 6 6 6 7 7 7 8 8 9 9
5 | 0 0 0 1 2 2 3 3 3 3 4 4 4 5 5 5 6 6 7 8 9
6 | 0 1 1 2 2 3 3 7 8 8
7 | 0
```

▲ Figure 2.6

If the data have been grouped and the original data have been lost, or are otherwise unknown, then a histogram may be considered. A grouped frequency table and histogram illustrating the marks are shown below.

Marks, x	Frequency, f
0–9	5
10–19	8
20–29	14
30–39	19
40–49	22
50–59	21
60–70	11

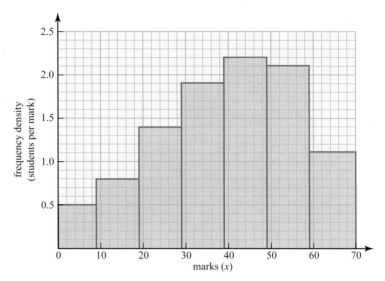

▲ Figure 2.7

> **Note**
>
> The class boundary 10–19 becomes $9.5 \leqslant x < 19.5$ for the purpose of drawing the histogram. You must give careful consideration to class boundaries, particularly if you are using rounded data.

> **?**
>
> ❯ Look at the intervals for the first and last classes. How do they differ from the others? Why is this the case?

Grouped discrete data are illustrated well by a histogram if the distribution is particularly skewed as is the case in the next example.

The first 50 positive integers squared are:

1	4	9	16	25	36	49	64
81	100	121	144	169	196	225	256
289	324	361	400	441	484	529	576
625	676	729	784	841	900	961	1024
1089	1156	1225	1296	1369	1444	1521	1600
1681	1764	1849	1936	2025	2116	2209	2304
2401	2500						

Number, n	Frequency, f
$0 < n \leqslant 250$	15
$250 < n \leqslant 500$	7
$500 < n \leqslant 750$	5
$750 < n \leqslant 1000$	4
$1000 < n \leqslant 1250$	4
$1250 < n \leqslant 1500$	3
$1500 < n \leqslant 1750$	3
$1750 < n \leqslant 2000$	3
$2000 < n \leqslant 2250$	3
$2250 < n \leqslant 2500$	3

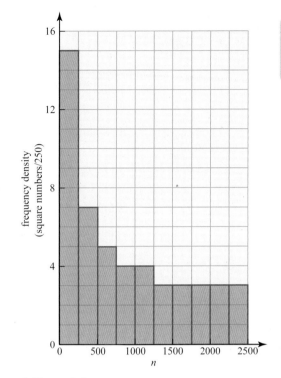

▲ Figure 2.8

The main points to remember when drawing a histogram are:

» Histograms are usually used for illustrating continuous data. For discrete data it is better to draw a stem-and-leaf diagram, line graph or bar chart.

» Since the data are continuous, or treated as if they were continuous, adjacent columns of the histogram should touch (unlike a bar chart where the columns should be drawn with gaps between them).

» It is the areas and not the heights of the columns that are proportional to the frequency of each class.

» The vertical axis should be marked with the appropriate frequency density (*frequency per 5 mm* for example), rather than frequency.

Exercise 2A

1 A number of trees in two woods were measured. Their diameters, correct to the nearest centimetre, are summarised in the table below.

Diameter (cm)	1–10	11–15	16–20	21–30	31–50	Total
Mensah's Wood	10	5	3	11	1	30
Ashanti Forest	6	8	20	5	1	40

(Trees less than 0.5 cm in diameter are not included.)

(i)   Write down the actual class boundaries.

(ii) Draw two separate histograms to illustrate this information.

(iii) State the modal class for each wood.

(iv) Describe the main features of the distributions for the two woods.

CP **2** Listed below are the prime numbers, p, from 1 up to 1000. (1 itself is not usually defined as a prime.)

Primes up to 1000

2	3	5	7	11	13	17	19	23	29	31	37	41	43
47	53	59	61	67	71	73	79	83	89	97	101	103	107
109	113	127	131	137	139	149	151	157	163	167	173	179	181
191	193	197	199	211	223	227	229	233	239	241	251	257	263
269	271	277	281	283	293	307	311	313	317	331	337	347	349
353	359	367	373	379	383	389	397	401	409	419	421	431	433
439	443	449	457	461	463	467	479	487	491	499	503	509	521
523	541	547	557	563	569	571	577	587	593	599	601	607	613
617	619	631	641	643	647	653	659	661	673	677	683	691	701
709	719	727	733	739	743	751	757	761	769	773	787	797	809
811	821	823	827	829	839	853	857	859	863	877	881	883	887
907	911	919	929	937	941	947	953	967	971	977	983	991	997

(i)   Draw a histogram to illustrate these data with the following class intervals:

$1 \leqslant p < 20$ $20 \leqslant p < 50$ $50 \leqslant p < 100$ $100 \leqslant p < 200$
$200 \leqslant p < 300$ $300 \leqslant p < 500$ and $500 \leqslant p < 1000$.

(ii) Comment on the shape of the distribution.

3 A crate containing 270 oranges was opened and each orange was weighed to the nearest gram. The masses were found to be distributed as in this table.

(i)   Draw a histogram to illustrate the data.

(ii) From the table, calculate an estimate of the mean mass of an orange from this crate.

Mass (grams)	Number of oranges
60–99	20
100–119	60
120–139	80
140–159	50
160–219	60

4   In an agricultural experiment, 320 plants were grown on a plot, and the lengths of the stems were measured to the nearest centimetre ten weeks after planting. The lengths were found to be distributed as in this table.

Length (cm)	Number of plants
20–31	30
32–37	80
38–43	90
44–49	60
50–67	60

(i)   Draw a histogram to illustrate the data.

(ii) From the table, calculate an estimate of the mean length of stem of a plant from this experiment.

5   The lengths of time of sixty songs recorded by a certain group of singers are summarised in the table below.

Song length in seconds (x)	Number of songs
$0 < x < 120$	1
$120 \leqslant x < 180$	9
$180 \leqslant x < 240$	15
$240 \leqslant x < 300$	17
$300 \leqslant x < 360$	13
$360 \leqslant x < 600$	5

(i)   Display the data on a histogram.

(ii) Determine the mean song length.

CP 6   The doctors in a hospital in an African country are concerned about the health of pregnant women in the region. They decide to record and monitor the weights of newborn babies.

(i)   A junior office clerk suggests recording the babies' weights in this frequency table.

Weight (kg)	2.0–2.5	2.5–3.0	3.0–3.5	3.5–4.0
Frequency	*Doctors*	*to*	*fill*	*in*

Give three criticisms of the table.

(ii) After some months the doctors have enough data to draw this histogram.

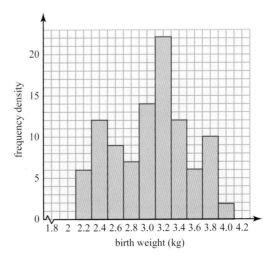

The doctors consider babies less than 2.7 kg to be small and those over 3.9 kg to be large.

They then consider a number of possible courses of action:

» carry on as at present

» induce any babies that are more than a week overdue

» set up a special clinic for 'at risk' pregnant women.

Comment on what features of the histogram give rise to each of these suggestions.

(iii) The doctors decide that the best option is to set up a special clinic. Suggest what other data they will need in order to make a strong case for funding.

7 After completing a long assignment, a student was told by his tutor that it was more like a book than an essay. He decided to investigate how many pages there are in a typical book and started by writing down the numbers of pages in the books on one of his shelves, as follows.

256	128	160	128	192	64	356	96	64	160
464	128	96	96	556	148	64	192	96	512
940	676	128	196	640	44	64	144	256	72

(i) Look carefully at the data and state, giving your reasons, whether they are continuous or discrete. Give an explanation for your answer.

(ii) Decide on the most helpful method of displaying the data and draw the appropriate diagram.

8 As part of a data collection exercise, members of a certain school year group were asked how long they spent on their Mathematics homework during one particular week. The times are given to the nearest 0.1 hour. The results are displayed in the following table.

Time spent (t hours)	$0.1 \leqslant t \leqslant 0.5$	$0.6 \leqslant t \leqslant 1.0$	$1.1 \leqslant t \leqslant 2.0$	$2.1 \leqslant t \leqslant 3.0$	$3.1 \leqslant t \leqslant 4.5$
Frequency	11	15	18	30	21

(i) Draw, on graph paper, a histogram to illustrate this information.

(ii) Calculate an estimate of the mean time spent on their Mathematics homework by members of this year group.

Cambridge International AS & A Level Mathematics
9709 Paper 6 Q5 June 2008

2.2 Measures of central tendency and of spread using quartiles

You saw in Chapter 1 how to find the median of a set of discrete data. As a reminder, the median is the value of the middle item when all the data items have been ranked in order.

The median is the value of the $\frac{n + 1}{2}$th item and is half-way through the data set.

The values one-quarter of the way through the data set and three-quarters of the way through the data set are called the **lower quartile** and the **upper quartile** respectively. The lower quartile, median and upper quartile are usually denoted using Q_1, Q_2 and Q_3.

Quartiles are used mainly with large data sets and their values found by looking at the $\frac{1}{4}$, $\frac{1}{2}$ and $\frac{3}{4}$ points. So, for a data set of 1000, you would take Q_1 to be the value of the 250th data item, Q_2 to be the value of the 500th data item and Q_3 to be the value of the 750th data item.

Quartiles for small data sets

For small data sets, where each data item is known (raw data), calculation of the middle quartile Q_2, the median, is straightforward. However, there are no standard formulae for the calculation of the lower and upper quartiles, Q_1 and Q_3, and you may meet different ones. The one we will use is consistent with the output from some calculators which display the quartiles of a data set and depends on whether the number of items, n, is even or odd.

If n is *even* then there will be an equal number of items in the lower half and upper half of the data set. To calculate the lower quartile, Q_1, find the median of the lower half of the data set. To calculate the upper quartile, Q_3, find the median of the upper half of the data set.

For example, for the data set $\{1, 3, 6, 10, 15, 21, 28, 36, 45, 55\}$ the median, Q_2, is $\frac{15 + 21}{2} = 18$. The lower quartile, Q_1, is the median of $\{1, 3, 6, 10, 15\}$, i.e. 6. The upper quartile, Q_3, is the median of $\{21, 28, 36, 45, 55\}$, i.e. 36.

If n is *odd* then define the 'lower half' to be all data items *below* the median. Similarly define the 'upper half' to be all data items *above* the median. Then proceed as if n were even.

For example, for the data set $\{1, 3, 6, 10, 15, 21, 28, 36, 45\}$ the median, Q_2, is 15. The lower quartile, Q_1, is the median of $\{1, 3, 6, 10\}$, i.e. $\frac{3 + 6}{2} = 4.5$. The upper quartile, Q_3, is the median of $\{21, 28, 36, 45\}$, i.e. $\frac{28 + 36}{2} = 32$.

> The definition of quartiles on a spreadsheet may be different from that described above. Values of Q_1 and Q_3 in the even case shown above are given as 7 and 34 respectively on an *Excel* spreadsheet. Similarly, values of Q_1 and Q_3 in the odd case shown above are given as 6 and 28 respectively.

> **ACTIVITY 2.1**
>
> If you have access to a computer, use a spreadsheet to find the median and quartiles of a small data set. Find out the method the spreadsheet uses to determine the position of the lower and upper quartiles.

Example 2.1

Catherine is a junior reporter. As part of an investigation into consumer affairs she purchases 0.5 kg of chicken from 12 shops and supermarkets in the town. The resulting data, put into rank order, are as follows:

$1.39 $1.39 $1.46 $1.48 $1.48 $1.50 $1.52 $1.54 $1.60 $1.65 $1.68 $1.72

Find Q_1, Q_2 and Q_3.

Solution

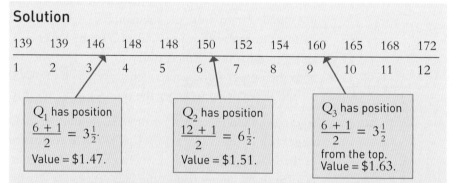

| 139 | 139 | 146 | 148 | 148 | 150 | 152 | 154 | 160 | 165 | 168 | 172 |
| 1 | 2 | 3 | 4 | 5 | 6 | 7 | 8 | 9 | 10 | 11 | 12 |

Q_1 has position
$$\frac{6+1}{2} = 3\tfrac{1}{2}.$$
Value = $1.47.

Q_2 has position
$$\frac{12+1}{2} = 6\tfrac{1}{2}.$$
Value = $1.51.

Q_3 has position
$$\frac{6+1}{2} = 3\tfrac{1}{2}$$
from the top.
Value = $1.63.

In fact, the upper quartile has a value of $1.625 but this has been rounded up to the nearest cent.

> **!**
>
> You may encounter different formulae for finding the lower and upper quartiles. The ones given here are relatively easy to calculate and usually lead to values of Q_1 and Q_3 which are close to the true values.

> **?**
>
> ❯ What are the true values?

Interquartile range or quartile spread

The difference between the lower and upper quartiles is known as the **interquartile range** or **quartile spread**.

➤ Interquartile range $(IQR) = Q_3 - Q_1$.

In Example 2.1, $IQR = 163 - 147 = 16$ cents.

The interquartile range covers the middle 50% of the data. It is relatively easy to calculate and is a useful measure of spread as it avoids extreme values. It is said to be resistant to outliers.

Box-and-whisker plots (boxplots)

The three quartiles and the two extreme values of a data set may be illustrated in a **box-and-whisker plot**. This is designed to give an easy-to-read representation of the location and spread of a distribution. Figure 2.9 shows a box-and-whisker plot for the data in Example 2.1.

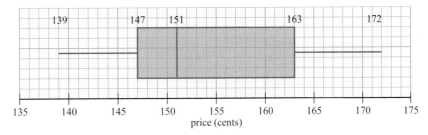

▲ **Figure 2.9**

The box represents the middle 50% of the distribution and the whiskers stretch out to the extreme values.

Outliers

In Chapter 1 you met a definition of an outlier based on the mean and standard deviation. A different approach gives the definition of an outlier in terms of the median and interquartile range (*IQR*).

Data that are more than $1.5 \times IQR$ beyond the lower or upper quartiles are regarded as outliers.

The corresponding boundary values beyond which outliers may be found are

$$Q_1 - 1.5 \times (Q_3 - Q_1) \quad \text{and} \quad Q_3 + 1.5 \times (Q_3 - Q_1).$$

For the data relating to the ages of the cyclists involved in accidents discussed in Chapter 1, for all 92 data values $Q_1 = 13.5$ and $Q_3 = 45.5$.

Hence $Q_1 - 1.5 \times (Q_3 - Q_1) = 13.5 - 1.5 \times (45.5 - 13.5)$

$$= 13.5 - 1.5 \times 32$$

$$= -34.5$$

and $Q_3 + 1.5 \times (Q_3 - Q_1) = 45.5 + 1.5 \times (45.5 - 13.5)$

$$= 45.5 + 1.5 \times 32$$

$$= 93.5$$

From these boundary values you will see that there are no outliers at the lower end of the range, but the value of 138 is an outlier at the upper end of the range.

Figure 2.10 shows a box-and-whisker plot for the ages of the cyclists with the outlier removed. For the remaining 91 data items $Q_1 = 13$ and $Q_3 = 45$.

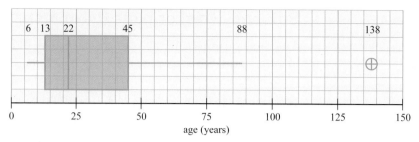

▲ **Figure 2.10**

From the diagram you can see that the distribution has positive or right skewness. The ⊕ indicates an outlier and is above the upper quartile. Outliers are usually labelled as they are often of special interest. The whiskers are drawn to the most extreme data points which are not outliers.

2.3 Cumulative frequency graphs

When working with large data sets or grouped data, percentiles and quartiles can be found from **cumulative frequency graphs** (also known as *cumulative frequency curves*) as shown in the next section.

Sheuligirl

I am a student trying to live on a small allowance. I'm trying my best to allow myself a sensible monthly budget but my lecturers have given me a long list of textbooks to buy. If I buy just half of them I will have nothing left to live on this month. The majority of books on my list are over $16.

I want to do well at my studies but I won't do well without books and I won't do well if I am ill through not eating properly.

Please tell me what to do, and don't say 'go to the library' because the books I need are never there.

After reading this opening post a journalist wondered if there was a story in it. He decided to carry out a survey of the prices of textbooks in a large shop. The reporter took a large sample of 470 textbooks and the results are summarised in the table.

Cost, C ($)	Frequency (no. of books)
$0 \leqslant C < 10$	13
$10 \leqslant C < 15$	53
$15 \leqslant C < 20$	97
$20 \leqslant C < 25$	145
$25 \leqslant C < 30$	81
$30 \leqslant C < 35$	40
$35 \leqslant C < 40$	23
$40 \leqslant C < 45$	12
$45 \leqslant C < 50$	6

He decided to estimate the median and the upper and lower quartiles of the costs of the books. (Without the original data you cannot find the actual values so all calculations will be estimates.) The first step is to make a cumulative frequency table, then to plot a cumulative frequency graph.

Cost, C ($)	Frequency	Cost	Cumulative frequency
$0 \leqslant C < 10$	13	$C < 10$	13
$10 \leqslant C < 15$	53	$C < 15$	66	← See note 1.
$15 \leqslant C < 20$	97	$C < 20$	163	← See note 2.
$20 \leqslant C < 25$	145	$C < 25$	308
$25 \leqslant C < 30$	81	$C < 30$	389
$30 \leqslant C < 35$	40	$C < 35$	429
$35 \leqslant C < 40$	23	$C < 40$	452
$40 \leqslant C < 45$	12	$C < 45$	464
$45 \leqslant C < 50$	6	$C < 50$	470

Notes

1 Notice that the interval $C < 15$ means $0 \leqslant C < 15$ and so includes the 13 books in the interval $0 \leqslant C < 10$ and the 53 books in the interval $10 \leqslant C < 15$, giving 66 books in total.

2 Similarly, to find the total for the interval $C < 20$ you must add the number of books in the interval $15 \leqslant C < 20$ to your previous total, giving you $66 + 97 = 163$.

A cumulative frequency graph is obtained by plotting the **upper boundary** of each class against the cumulative frequency. The points are joined by a smooth curve, as shown in Figure 2.11.

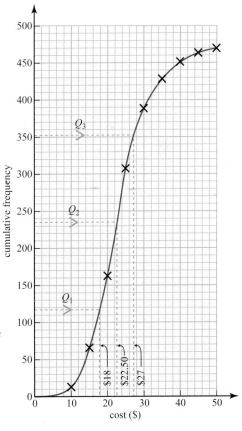

▲ Figure 2.11

> In this example the actual values are unknown and the median must therefore be an estimate. It is usual in such cases to find the *estimated* value of the $\frac{n}{2}$th item. This gives a better estimate of the median than is obtained by using $\frac{n+1}{2}$, which is used for ungrouped data. Similarly, estimates of the lower and upper quartiles are found from the $\frac{n}{4}$th and $\frac{3n}{4}$th items.

The 235th $\left(\frac{470}{2}\right)$ item of data identifies the median, which has a value of about $22.50. The 117.5th $\left(\frac{470}{2}\right)$ item of data identifies the lower quartile, which has a value of about $18 and the 352.5th $\left(\frac{3}{4} \times 470\right)$ item of data identifies the upper quartile, which has a value of about $27.

Notice the distinctive shape of the cumulative frequency graph. It is like a stretched-out S-shape leaning forwards.

What about Sheuligirl's claim that the majority of textbooks cost more than $16? $Q_1 = 18. By definition 75% of books are more expensive than this, so Sheuligirl's claim seems to be well founded. We need to check exactly how many books are estimated to be more expensive than $16.

From the cumulative frequency graph 85 books cost $16 or less (Figure 2.12). So 385 books, or about 82%, are more expensive.

> You should be cautious about any conclusions you draw. This example deals with books, many of which have prices like $9.95 or $39.99. In using a cumulative frequency graph you are assuming an even spread of data throughout the intervals and this may not always be the case.

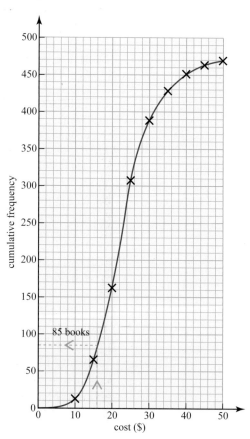

▲ Figure 2.12

Box-and-whisker plots for grouped data

It is often helpful to draw a box-and-whisker plot. In cases when the extreme values are unknown the whiskers are drawn out to the 10th and 90th percentiles. Arrows indicate that the minimum and maximum values are further out.

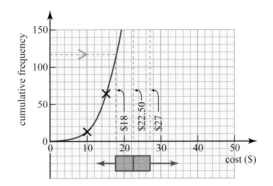

▲ **Figure 2.13**

| Example 2.2 |

A random sample of people were asked how old they were when they first met their partner. The histogram represents this information.

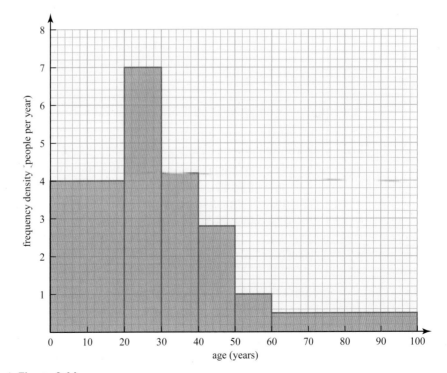

▲ **Figure 2.14**

(i)   What is the modal age group?

(ii) How many people took part in the survey?

(iii) Find an estimate for the mean age that a person first met their partner.

(iv) Draw a cumulative frequency graph for the data and use the graph to provide an estimate for the median.

Solution

(i)   The bar with the greatest frequency density represents the modal age group. So the modal age group is $20 \leqslant a < 30$.

(ii) Frequency density $= \dfrac{\text{frequency}}{\text{class width}}$

So,

$$\text{Frequency} = \text{frequency density} \times \text{class width}$$

Age (years)	Frequency density	Class width	Frequency
$0 \leqslant a < 20$	4	20	$4 \times 20 = 80$
$20 \leqslant a < 30$	7	10	$7 \times 10 = 70$
$30 \leqslant a < 40$	4.2	10	$4.2 \times 10 = 42$
$40 \leqslant a < 50$	2.8	10	$2.8 \times 10 = 28$
$50 \leqslant a < 60$	1	10	$1 \times 10 = 10$
$60 \leqslant a < 100$	0.5	40	$0.5 \times 40 = 20$

The total number of people is $80 + 70 + 42 + 28 + 10 + 20 = 250$.

> You will see that the class with the greatest frequency is $0 \leqslant a < 20$, with 80 people. However, this is not the modal class because its frequency density of 4 people per year is lower than the frequency density of 7 people per year for the $20 \leqslant a < 30$ class. The modal class is that with the highest frequency density. The class width for $0 \leqslant a < 20$ is twice that for $20 \leqslant a < 30$ and this is taken into account in working out the frequency density.

(iii) To find an estimate for the mean, work out the midpoint of each class multiplied by its frequency; then sum the results and divide the answer by the total frequency.

$$\text{Estimated mean} = \frac{80 \times 10 + 70 \times 25 + 42 \times 35 + 28 \times 45 + 10 \times 55 + 20 \times 80}{250}$$

$$= \frac{7430}{250}$$

$$= 29.7 \text{ years to 3 s.f.}$$

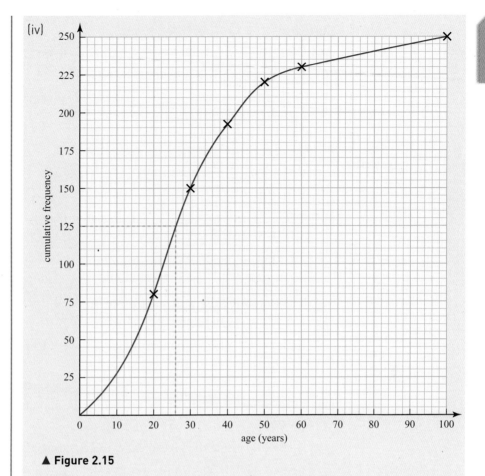

▲ Figure 2.15

The median age is 26 years.

| Example 2.3 | These are the times, in seconds, that 15 members of an athletics club took to run 800 metres. |

139   148   151   140   162

182   154   171   157   142

145   178   132   148   166

(i)   Draw a stem-and-leaf diagram of the data.

(ii) Find the median, the upper and lower quartiles and the interquartile range.

(iii) Draw a box-and-whisker plot of the data.

→

Solution

(i) $n = 15$

13 | 2 represents 132 seconds

```
13 | 2 9
14 | 0 2 5 8 8
15 | 1 4 7
16 | 2 6
17 | 1 8
18 | 2
```

(ii) There are 15 data values, so the median is the 8th data value.
So the median is 151 seconds.

The upper quartile is the median of the upper half of the data set.
So the upper quartile is 166 seconds.

The lower quartile is the median of the lower half of the data set.
So the lower quartile is 142 seconds.

$$\begin{aligned} \text{Interquartile range} &= \text{upper quartile} - \text{lower quartile} \\ &= 166 - 142 \\ &= 24 \text{ seconds} \end{aligned}$$

(iii) Draw a box that starts at the lower quartile and ends at the upper quartile.

Add a line inside the box to show the position of the median.

Extend the whiskers to the greatest and least values in the data set.

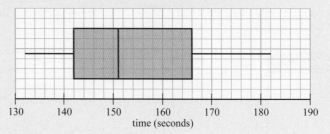

▲ Figure 2.16

Exercise 2B

1 For each of the following data sets, find:

(a) the range

(b) the median

(c) the lower and upper quartiles

(d) the interquartile range

(e) any outliers.

(i) 6   8   3   2   1   5   4   6   8   5   6   7   8   8   6   6

(ii) 12 5   17 11 4   10 12 19 12 5   9   15 11 16 8

     18 12 8   9   11   12 14 8   14 7

(iii) 25 28 33 14 37   19 23 27 25 28

(iv) 115 123 132 109 127   116 141 132 114 109

     125 121 117 118 117   116 123 105 125

2 (i) For the following data set, find the median and interquartile range.

     2   8   4   6   3   5   1   8   2   5   8   0   3   7   8   5

Use your answers to part (i) to deduce the median and interquartile range for each of the following data sets.

(ii) 32 38 34 36 33   35 31 38 32 35   38 30 33 37 38 35

(iii) 20 80 40 60 30   50 10 80 20 50   80 0   30 70 80 50

(iv) 50 110 70 90 60   80 40 110 50 80   110 30 60 100 110 80

CP **3** Find:

(i)   the median

(ii) the upper and lower quartiles

(iii) the interquartile range

for the scores of golfers in the first round of a competition.

Score	Tally
70	I
71	II
72	IIII
73	⊞⊞ III
74	⊞⊞ ⊞⊞ II
75	⊞⊞ II
76	⊞⊞
77	⊞⊞ I
78	
79	III
80	I
81	
82	I

(iv) Illustrate the data with a box-and-whisker plot.

(v) The scores for the second round are illustrated on the box-and-whisker plot below. Compare the two and say why you think the differences might have arisen.

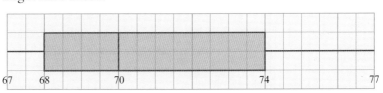

73

CP 4 The numbers of goals scored by a hockey team in its matches one season are illustrated on the vertical line chart below.

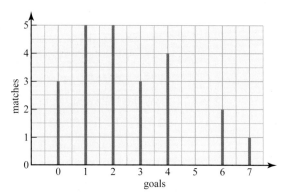

(i) Draw a box-and-whisker plot to illustrate the same data.

(ii) State, with reasons, which you think is the better method of display in this case.

CP 5 One year the yields, y, of a number of walnut trees were recorded to the nearest kilogram, as follows.

Yield, y (kg)	Frequency
$40 < y \leqslant 50$	1
$50 < y \leqslant 60$	5
$60 < y \leqslant 70$	7
$70 < y \leqslant 80$	4
$80 < y \leqslant 90$	2
$90 < y \leqslant 100$	1

(i) Construct the cumulative frequency table for these data.

(ii) Draw the cumulative frequency graph.

(iii) Use your graph to estimate the median and interquartile range of the yields.

(iv) Draw a box-and-whisker plot to illustrate the data.

The piece of paper where the actual figures had been recorded was then found, and these were:

44 59 67 76 52   62 68 78 53 63

85 93 56 65 74   69 82 53 65 70

(v) Use these data to find the median and interquartile range and compare your answers with those you obtained from the grouped data.

(vi) What are the advantages and disadvantages of grouping data?

6   The intervals of time, t seconds, between successive emissions from a weak radioactive source were measured for 200 consecutive intervals, with the following results.

Interval (t seconds)	$0 < t \leqslant 5$	$5 < t \leqslant 10$	$10 < t \leqslant 15$	$15 < t \leqslant 20$
Frequency	23	67	42	26
Interval (t seconds)	$20 < t \leqslant 25$	$25 < t \leqslant 30$	$30 < t \leqslant 35$	
Frequency	21	15	6	

(i)   Draw a cumulative frequency graph for this distribution.

(ii) Use your graph to estimate:

   (a)   the median

   (b)   the interquartile range.

(iii) Calculate an estimate of the mean of the distribution.

7   In a sample of 800 eggs from an egg farm each egg was weighed and classified according to its mass, m grams. The frequency distribution was as follows.

Mass in grams	$40 < m \leqslant 45$	$45 < m \leqslant 50$	$50 < m \leqslant 55$
Number of eggs	36	142	286
Mass in grams	$55 < m \leqslant 60$	$60 < m \leqslant 65$	$65 < m \leqslant 70$
Number of eggs	238	76	22

Draw a cumulative frequency graph of the data, using a scale of 2 cm to represent 5 grams on the horizontal axis (which should be labelled from 40 to 70 grams) and a scale of 2 cm to represent 100 eggs on the vertical axis. Use your graph to estimate for this sample:

(i)   the percentage of eggs that would be classified as large (over 62 grams)

(ii) the median mass of an egg

(iii) the interquartile range.

Indicate clearly on your diagram how you arrive at your results.

8   The table summarises the observed lifetimes, x, in seconds, of 50 fruit flies subjected to a new spray in a controlled experiment.

Interval	Mid-interval value	Frequency
$0.5 \leqslant x < 5.5$	3	3
$5.5 \leqslant x < 10.5$	8	22
$10.5 \leqslant x < 15.5$	13	12
$15.5 \leqslant x < 20.5$	18	9
$20.5 \leqslant x < 25.5$	23	2
$25.5 \leqslant x < 30.5$	28	1
$30.5 \leqslant x < 35.5$	33	1

CP

(i) Making clear your methods and showing all your working, estimate the mean and standard deviation of these lifetimes. Give your answers correct to 3 significant figures and do not make any corrections for grouping.

(ii) Draw the cumulative frequency graph and use it to estimate the minimum lifetime below which 70% of all lifetimes lie.

9 During January the numbers of people entering a store during the first hour after opening were as follows.

Time after opening, x minutes	Frequency	Cumulative frequency
$0 < x \leqslant 10$	210	210
$10 < x \leqslant 20$	134	344
$20 < x \leqslant 30$	78	422
$30 < x \leqslant 40$	72	a
$40 < x \leqslant 60$	b	540

(i) Find the values of a and b.

(ii) Draw a cumulative frequency graph to represent this information. Take a scale of 2 cm for 10 minutes on the horizontal axis and 2 cm for 50 people on the vertical axis.

(iii) Use your graph to estimate the median time after opening that people entered the store.

(iv) Calculate estimates of the mean, m minutes, and standard deviation, s minutes, of the time after opening that people entered the store.

(v) Use your graph to estimate the number of people entering the store between $(m - \frac{1}{2}s)$ and $(m + \frac{1}{2}s)$ minutes after opening.

Cambridge International AS & A Level Mathematics
9709 Paper 6 Q6 June 2009

10 The numbers of people travelling on a certain bus at different times of the day are as follows.

17   5   2   23   16   31   8
22   14   25   35   17   27   12
6   23   19   21   23   8   26

(i) Draw a stem-and-leaf diagram to illustrate the information given above.

(ii) Find the median, the lower quartile, the upper quartile and the interquartile range.

(iii) State, in this case, which of the median and mode is preferable as a measure of central tendency, and why.

Cambridge International AS & A Level Mathematics
9709 Paper 61 Q2 June 2010

1 Histograms:
  - commonly used to illustrate continuous data
  - horizontal axis shows the variable being measured (cm, kg, etc.)
  - vertical axis labelled frequency density where
    $$\text{frequency density} = \frac{\text{frequency}}{\text{class width}}$$
  - no gaps between columns
  - the frequency is proportional to the *area* of each column.

2 For a small data set with n items,
  - the median, Q_2, is the value of the $\frac{n+1}{2}$ th item of data.
    If n is even then:
  - the lower quartile, Q_1, is the median of the lower half of the data set
  - the upper quartile, Q_3, is the median of the upper half of the data set.

  If n is odd then exclude the median from either 'half' and proceed as if n were even.

3 Interquartile range $(IQR) = Q_3 - Q_1$.

4 When data are illustrated using a cumulative frequency graph, the median and the lower and upper quartiles are estimated by identifying the data values with cumulative frequencies $\frac{1}{2}n$, $\frac{1}{4}n$ and $\frac{3}{4}n$.

5 An item of data x may be identified as an *outlier* if it is more than $1.5 \times IQR$ beyond the lower or upper quartile, i.e. if
  $$x < Q_1 - 1.5 \times (Q_3 - Q_1) \quad \text{or} \quad x > Q_3 + 1.5 \times (Q_3 - Q_1).$$

6 A box-and-whisker plot is a useful way of summarising data and showing the median, upper and lower quartiles and any outliers.

LEARNING OUTCOMES

Now that you have finished this chapter, you should be able to
- draw the following diagrams:
  - histogram
  - box-and-whisker plot
  - cumulative frequency graph
- select a suitable way of presenting raw statistical data
- discuss advantages and/or disadvantages of a particular way of presenting data
- use a cumulative frequency graph to calculate or estimate the:
  - lower quartile
  - upper quartile
  - interquartile range
  - proportion in a given interval
- identify outliers.

3

Probability

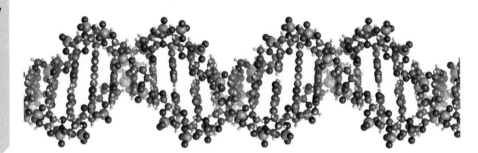

The Librarian Blogosphere

A library without books

If you plan to pop into your local library and pick up the latest bestseller, then forget it. All the best books 'disappear' practically as soon as they are put on the shelves.

I talked about the problem with the local senior librarian, Gina Clarke.

'We have a real problem with unauthorised loans at the moment,' Gina told me. 'Out of our total stock of, say, 80 000 books, something like 44 000 are out on loan at any one time. About 20 000 are on the shelves and I'm afraid the rest are unaccounted for.'

Librarian Gina Clarke is worried about the problem of 'disappearing books'

That means that the probability of finding a particular book you want from the library's list is exactly $\frac{1}{4}$. With odds like that, don't count on being lucky next time you visit your library.

❓

> How do you think the figure of $\frac{1}{4}$ at the end of the article was arrived at?
> Do you agree that the probability is *exactly* $\frac{1}{4}$?

The information about the different categories of book can be summarised as follows.

Category of book	Typical numbers
On the shelves	20 000
Out on loan	44 000
Unauthorised loan	16 000
Total stock	**80 000**

On the basis of these figures it is possible to estimate the probability of finding the book you want. Of the total stock of 80 000 books bought by the library, you might expect to find about 20 000 on the shelves at any one time. As a fraction, this is $\frac{20}{80}$ or $\frac{1}{4}$ of the total. So, as a rough estimate, the probability of your finding a particular book is 0.25 or 25%.

Similarly, 16 000 out of the total of 80 000 books are on unauthorised loan, a euphemism for *stolen*, and this is 20%, or $\frac{1}{5}$.

An important assumption underlying these calculations is that all the books are equally likely to be unavailable, which is not very realistic since popular books are more likely to be stolen. Also, the numbers given are only rough approximations, so it is definitely incorrect to say that the probability is *exactly* $\frac{1}{4}$.

3.1 Measuring probability

Probability (or chance) is a way of describing the likelihood of different possible **outcomes** occurring as a result of some **experiment**.

In the example of the library books, the experiment is looking in the library for a particular book. Let us assume that you already know that the book you want is on the library's stocks. The three possible outcomes are that the book is on the shelves, out on loan or missing.

It is important in probability to distinguish experiments from the outcomes that they may generate. A list of all possible outcomes is called a **sample space**. Here are a few examples.

Experiment	Possible outcomes
Guessing the answer to a four-option multiple choice question	A B C D
Predicting the next vehicle to go past the corner of my road	car bus lorry bicycle van other
Tossing a coin	heads tails

Another word for experiment is **trial**. This is used in Chapter 6 of this book to describe the binomial situation where there are just two possible outcomes.

Another word you should know is **event**. This often describes several outcomes put together. For example, when rolling a die, an event could be 'the die shows an even number'. This event corresponds to three different outcomes from the trial, the die showing 2, 4 or 6. However, the term *event* is also often used to describe a single outcome.

3.2 Estimating probability

Probability is a number that measures likelihood. It may be estimated experimentally or theoretically.

Experimental estimation of probability

In many situations probabilities are estimated on the basis of data collected experimentally, as in the following example.

Of 30 drawing pins tossed in the air, 21 of them were found to have landed with their pins pointing up. From this you would estimate the probability that the next pin tossed in the air will land with its pin pointing up to be $\frac{21}{30}$ or 0.7.

You can describe this in more formal notation.

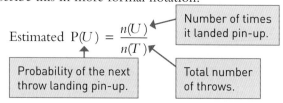

Theoretical estimation of probability

There are, however, some situations where you do not need to collect data to make an estimate of probability.

For example, when tossing a coin, common sense tells you that there are only two realistic outcomes and, given the symmetry of the coin, you would expect them to be equally likely. So the probability, P(H), that the next coin will produce the outcome heads can be written as follows:

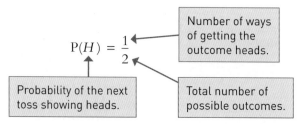

Example 3.1

Using the notation described above, write down the probability that the correct answer for the next four-option multiple choice question will be answer A. What assumptions are you making?

Solution

Assuming that the test-setter has used each letter equally often, the probability, $P(A)$, that the next question will have answer A can be written as follows:

$$P(A) = \frac{1}{4}$$

Answer A.

Answers A, B, C and D.

Notice that we have assumed that the four options are equally likely. Equiprobability is an important assumption underlying most work on probability.

Expressed formally, the probability, $P(A)$, of event A occurring is:

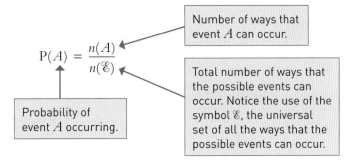

$$P(A) = \frac{n(A)}{n(\mathscr{E})}$$

Number of ways that event A can occur.

Probability of event A occurring.

Total number of ways that the possible events can occur. Notice the use of the symbol $\mathscr{E}$, the universal set of all the ways that the possible events can occur.

Probabilities of 0 and 1

The two extremes of probability are **certainty** at one end of the scale and impossibility at the other. Here are examples of certain and impossible events.

Experiment	Certain event	Impossible event
Rolling a single die	The result is in the range 1 to 6 inclusive	The result is a 7
Tossing a coin	Getting either heads or tails	Getting neither heads nor tails

Certainty

As you can see from the table above, for events that are certain, the number of ways that the event can occur, $n(A)$ in the formula, is equal to the total number of possible events, $n(\mathscr{E})$.

$$\frac{n(A)}{n(\mathscr{E})} = 1$$

So the probability of an event that is certain is one.

Impossibility

For impossible events, the number of ways that the event can occur, $n(A)$, is zero.

$$\frac{n(A)}{n(\mathscr{E})} = \frac{0}{n(\mathscr{E})} = 0$$

So the probability of an event that is impossible is zero.

Typical values of probabilities might be something like 0.3 or 0.9. If you arrive at probability values of, say, −0.4 or 1.7, you will know that you have made a mistake since these are meaningless.

$$0 \leqslant P(A) \leqslant 1 \longleftarrow \boxed{\text{Certain event.}}$$

$$\boxed{\text{Impossible event.}}$$

The complement of an event

The complement of an event A, denoted by A', is the event *not-A*, that is the event 'A does not happen'.

Example 3.2

It was found that, out of a box of 50 matches, 45 lit but the others did not. What was the probability that a randomly selected match would not have lit?

Solution

The probability that a randomly selected match lit was

$$P(A') = \frac{45}{50} = 0.9.$$

The probability that a randomly selected match did not light was

$$P(A') = \frac{50 - 45}{50} = \frac{5}{50} = 0.1.$$

From this example you can see that

$$P(A') = 1 - P(A)$$

$$\boxed{\text{The probability of } A \text{ not occurring.}} \qquad \boxed{\text{The probability of } A \text{ occurring.}}$$

This is illustrated in Figure 3.1.

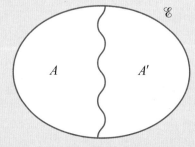

▲ **Figure 3.1** Venn diagram showing events A and A' (i.e. *not-A*)

3.3 Expectation

Health services braced for flu epidemic

Local health services are poised for their biggest challenge in years. The virulent strain of flu, named Trengganu B from its origins in Malaysia, currently sweeping across the world, is expected to hit any day.

With a chance of one in three of any individual contracting the disease, and 120 000 people within the Health Area, surgeries and hospitals are expecting to be swamped with patients.

Local doctor Aloke Ghosh says 'Immunisation seems to be ineffective against this strain'.

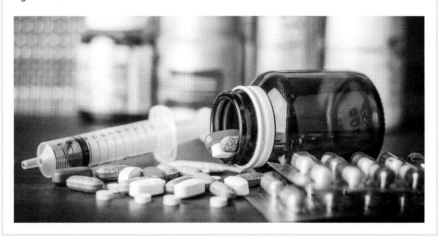

How many people can the health services expect to contract flu? The answer is easily seen to be $120\,000 \times \frac{1}{3} = 40\,000$. This is called the **expectation** or **expected frequency** and is given in this case by np, where n is the population size and p the probability.

Expectation is a technical term and need not be a whole number. Thus the expectation of the number of heads when a coin is tossed 5 times is $5 \times \frac{1}{2} = 2.5$. You would be wrong to go on to say 'That means either 2 or 3' or to qualify your answer as 'about $2\frac{1}{2}$'. The expectation is 2.5.

The idea of expectation of a discrete random variable is explored more thoroughly in Chapter 4. Applications of the binomial distribution are covered in Chapter 6.

3.4 The probability of either one event or another

So far we have looked at just one event at a time. However, it is often useful to bracket two or more of the events together and calculate their combined probability.

Example 3.3

The table below is based on the data at the beginning of this chapter and shows the probability of the next book requested falling into each of the three categories listed, assuming that each book is equally likely to be requested.

Category of book	Typical numbers	Probability
On the shelves (S)	20 000	0.25
Out on loan (L)	44 000	0.55
Unauthorised loan (U)	16 000	0.20
Total ($S + L + U$)	**80 000**	**1**

What is the probability that a randomly requested book is *either* out on loan *or* on unauthorised loan (i.e. that it is not available)?

Solution

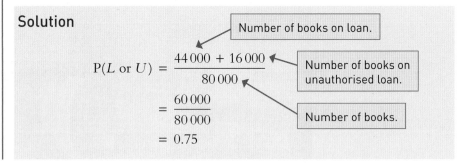

$$P(L \text{ or } U) = \frac{44\,000 + 16\,000}{80\,000}$$

$$= \frac{60\,000}{80\,000}$$

$$= 0.75$$

This can be written in more formal notation as

$$P(L \cup U) = \frac{n(L \cup U)}{n(\mathscr{E})}$$

$$= \frac{n(L)}{n(\mathscr{E})} + \frac{n(U)}{n(\mathscr{E})}$$

$$P(L \cup U) = P(L) + P(U)$$

Notice the use of the *union* symbol, $\cup$, to mean *or*. This is illustrated in Figure 3.2.

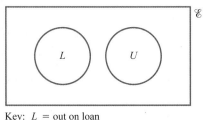

Key: L = out on loan
U = out on unauthorised loan

▲ **Figure 3.2** Venn diagram showing events L and U. It is not possible for both to occur.

In this example you could add the probabilities of the two events to get the combined probability of *either one or the other* event occurring. However, you have to be very careful adding probabilities as you will see in the next example.

Example 3.4

3

Below are further details of the categories of books in the library.

Category of book	Number of books
On the shelves	20 000
Out on loan	44 000
Adult fiction	22 000
Adult non–fiction	40 000
Junior	18 000
Unauthorised loan	16 000
Total stock	**80 000**

Asaph is trying to find the probability that the next book requested will be either out on loan or a book of adult non–fiction.

He writes:

Assuming all the books in the library are equally likely to be requested,

$$P(\text{on loan}) + P(\text{adult non-fiction}) = \frac{44\,000}{80\,000} + \frac{40\,000}{80\,000}$$

$$= 0.55 + 0.5$$

$$= 1.05$$

Explain why Asaph's answer must be wrong. What is his mistake?

Solution

This answer is clearly wrong as you cannot have a probability greater than 1.

The way this calculation was carried out involved some double counting. Some of the books classed as adult non-fiction were counted twice because they were also in the on-loan category, as you can see from Figure 3.3.

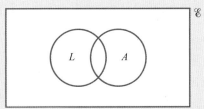

Key: L = out on loan
A = adult non-fiction

▲ **Figure 3.3** Venn diagram showing events L and A. It is possible for both to occur.

If you add all six of the book categories together, you find that they add up to 160 000, which represents twice the total number of books owned by the library.

A more useful representation of the data in the previous example is given in the two-way table below.

	Adult fiction	Adult non-fiction	Junior	Total
On the shelves	4 000	12 000	4 000	20 000
Out on loan	14 000	20 000	10 000	44 000
Unauthorised loan	4 000	8 000	4 000	16 000
Totals	22 000	40 000	18 000	80 000

If you simply add 44 000 and 40 000, you *double count* the 20 000 books that fall into both categories. So you need to subtract the 20 000 to ensure that it is counted only once. Thus:

Number either out on loan or adult non-fiction

$$= 44\,000 + 40\,000 - 20\,000$$
$$= 64\,000 \text{ books.}$$

So, the required probability $= \dfrac{64\,000}{80\,000} = 0.8.$

Mutually exclusive events

The problem of double counting does not occur when adding two rows in the table. Two rows cannot overlap, or **intersect**, which means that those categories are **mutually exclusive** (i.e. the one excludes the other, also known as **exclusive**). The same is true for two columns within the table.

Where two events, A and B, are mutually exclusive, the probability that either A or B occurs is equal to the sum of the separate probabilities of A and B occurring.

Where two events, A and B, are *not* mutually exclusive, the probability that either A or B occurs is equal to the sum of the separate probabilities of A and B occurring minus the probability of A and B occurring together.

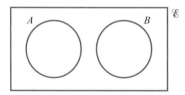   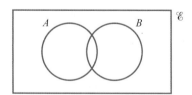

▲ **Figure 3.4** (a) Mutually exclusive events   (b) Not mutually exclusive events

$\text{P}(A \text{ or } B) = \text{P}(A) + \text{P}(B)$
$\text{P}(A \cup B) = \text{P}(A) + \text{P}(B)$

$\text{P}(A \text{ or } B) = \text{P}(A) + \text{P}(B) - \text{P}(A \text{ and } B)$
$\text{P}(A \cup B) = \text{P}(A) + \text{P}(B) - \text{P}(A \cap B)$

> Notice the use of the intersection sign, ∩, to mean *both ... and ...*

3

| Example 3.5 | A fair die is thrown. What is the probability that it shows each of these? |

(i)   Event A: an even number

(ii) Event B: a number greater than 4

(iii) Either A or B (or both): a number which is either even or greater than 4

Solution

(i)   Event A:
      Three out of the six numbers on a die are even, namely 2, 4 and 6.

      So $\quad P(A) = \frac{3}{6} = \frac{1}{2}$.

(ii) Event B:
      Two out of the six numbers on a die are greater than 4, namely 5 and 6.

      So $\quad P(B) = \frac{2}{6} = \frac{1}{3}$.

(iii) Either A or B (or both):
      Four of the numbers on a die are either even or greater than 4, namely 2, 4, 5 and 6.

      So $\quad P(A \cup B) = \frac{4}{6} = \frac{2}{3}$.

      This could also be found using

$$P(A \cup B) = P(A) + P(B) - P(A \cap B)$$

$$P(A \cup B) = \frac{3}{6} + \frac{2}{6} - \frac{1}{6}$$

$$= \frac{4}{6} = \frac{2}{3}$$

> This is the number 6 which is both even and greater than 4.

Exercise 3A	

1 Three separate electrical components, switch, bulb and contact point, are used together in the construction of a pocket torch. Of 534 defective torches, examined to identify the cause of failure, 468 are found to have a defective bulb. For a given failure of the torch, what is the probability that either the switch or the contact point is responsible for the failure? State clearly any assumptions that you have made in making this calculation.

2 If a fair die is thrown, what is the probability that it shows:

(i)   4

(ii) 4 or more

(iii) less than 4

(iv) an even number?

3   A bag containing *Scrabble* letters has the following letter distribution.

A	B	C	D	E	F	G	H	I	J	K	L	M
9	2	2	4	12	2	3	2	9	1	1	4	2

N	O	P	Q	R	S	T	U	V	W	X	Y	Z
6	8	2	1	6	4	6	4	2	2	1	2	1

The first letter is chosen at random from the bag; find the probability that it is:

(i)   an E

(ii)   in the first half of the alphabet

(iii) in the second half of the alphabet

(iv)   a vowel

(v)   a consonant

(vi)   the only one of its kind.

CP

4   A sporting chance

(i)   Two players, *A* and *B*, play tennis. On the basis of their previous results, the probability of *A* winning, P(*A*), is calculated to be 0.65. What is P(*B*), the probability of *B* winning?

(ii)   Two hockey teams, *A* and *B*, play a game. On the basis of their previous results, the probability of team *A* winning, P(*A*), is calculated to be 0.65. Why is it not possible to calculate directly P(*B*), the probability of team *B* winning, without further information?

(iii) In a tennis tournament, player *A*, the favourite, is estimated to have a 0.3 chance of winning the competition. Player *B* is estimated to have a 0.15 chance. Find the probability that either *A* or *B* will win the competition.

(iv)   In the Six Nations Rugby Championship, France and England are each given a 25% chance of winning or sharing the championship cup. It is also estimated that there is a 5% chance that they will share the cup. Estimate the probability that either England or France will win or share the cup.

5   The diagram shows even (*E*), odd (*O*) and square (*S*) numbers.

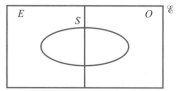

(i)   Copy the diagram and place the numbers 1 to 20 on it.

The numbers 1 to 20 are written on separate cards.

(ii)   A card is chosen at random. Find the probability that the number showing is:

(a)   even, *E*

(b)   square, *S*

(c)   odd, *O*

(d)   both even and square, $E \cap S$

(e)   either even or square, $E \cup S$

(f)   both even and odd, $E \cap O$

(g)   either even or odd, $E \cup O$.

Write down equations connecting the probabilities of the following events.

(h) $E, S, E \cap S, E \cup S$

(i) $E, O, E \cap O, E \cup O$

3.5 Independent and dependent events

My lucky day!

Won $100 when the number on my newspaper came up in the daily draw and $50 in the weekly draw too. A chance in a million!

Veronica

This story describes two pieces of good fortune on the same day. Veronica said the probability was about $\frac{1}{1\,000\,000}$. What was it really?

The two events resulted from two different experiments, the daily draw and the weekly draw. Consequently this situation is different from those you met in the previous section. There you were looking at two events from a single experiment (like the number coming up when a die is thrown being even or being greater than 4).

The total number of entrants in the daily draw was 1245 and in the weekly draw 324. The draws were conducted fairly, that is each number had an equal chance of being selected. The following table sets out the two experiments and their corresponding events with associated probabilities.

Experiment	Events (and estimated probabilities)
Daily draw	Winning: $\frac{1}{1245}$
	Not winning: $\frac{1244}{1245}$
Weekly draw	Winning: $\frac{1}{324}$
	Not winning: $\frac{323}{324}$

The two events 'win daily draw' and 'win weekly draw' are **independent events**. Two events are said to be independent when the outcome of the first event does not affect the outcome of the second event. The fact that Veronica has won the daily draw does not alter her chances of winning the weekly draw.

▸▸   For two independent events, A and B, $P(A \cap B) = P(A) \times P(B)$.

In situations like this the possible outcomes resulting from the different experiments are often shown on a **tree diagram**.

Example 3.6

Find, in advance of the results of the two draws, the probability that:

(i)   Veronica would win both draws

(ii) Veronica would fail to win either draw

(iii) Veronica would win one of the two draws.

Solution

The possible results are shown on the tree diagram in Figure 3.5.

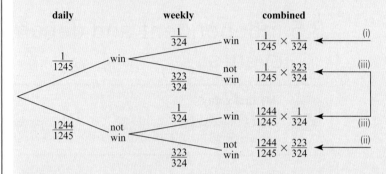

▲ **Figure 3.5**

(i)   The probability that Veronica wins both

$$= \frac{1}{1245} \times \frac{1}{324} = \frac{1}{403\,380}$$

This is not quite Veronica's 'one in a million' but it is not very far off it.

(ii) The probability that Veronica wins neither

$$= \frac{1244}{1245} \times \frac{323}{324} = \frac{401\,812}{403\,380}$$

This of course is much the most likely outcome.

(iii) The probability that Veronica wins one but not the other

$$= \underbrace{\frac{1}{1245} \times \frac{323}{324}}_{} + \underbrace{\frac{1244}{1245} \times \frac{1}{324}}_{} = \frac{1567}{403\,380}$$

> Wins daily draw but not weekly draw.

> Wins weekly draw but not daily draw.

Look again at the structure of the tree diagram in Figure 3.5.

There are two experiments, the daily draw and the weekly draw. These are considered as **First, Then** experiments, and set out *First* on the left and *Then* on the right. Once you understand this, the rest of the layout falls into place, with the different outcomes or events appearing as branches. In this example there are two branches at each stage; sometimes there may be three or more.

Notice that for a given situation the component probabilities sum to 1, as before.

$$\frac{1}{403\,380} + \frac{323}{403\,380} + \frac{1244}{403\,380} + \frac{401\,812}{403\,380} = \frac{403\,380}{403\,380} = 1$$

Example 3.7

Some friends buy a six-pack of popcorn. Two of the bags are salted (*S*), the rest are fruit flavoured (*F*). They decide to allocate the bags by lucky dip.

Find the probability that:

(i)   the first two bags chosen are the same as each other

(ii) the first two bags chosen are different from each other.

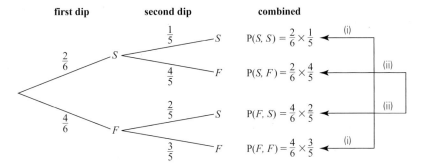

▲ **Figure 3.6**

Solution

Note: P(*F*, *S*) means the probability of drawing a fruit-flavoured bag (*F*) on the first dip and a salted bag (*S*) on the second.

(i)   The probability that the first two bags chosen are the same as each other is

$$P(S, S) + P(F, F) = \frac{2}{6} \times \frac{1}{5} + \frac{4}{6} \times \frac{3}{5}$$

$$= \frac{1}{15} + \frac{6}{15}$$

$$= \frac{7}{15}$$

(ii) The probability that the first two bags chosen are different from each other is

$$P(S, F) + P(F, S) = \frac{2}{6} \times \frac{4}{5} + \frac{4}{6} \times \frac{2}{5}$$

$$= \frac{4}{15} + \frac{4}{15}$$

$$= \frac{8}{15}$$

Note

The answer to part (ii) hinged on the fact that two orderings (*S* then *F*, and *F* then *S*) are possible for the same combined event (that the two bags selected include one salted and one fruit-flavoured bag).

The probabilities changed between the first dip and the second dip. This is because the outcome of the second dip is **dependent** on the outcome of the first one (with fewer bags remaining to choose from).

By contrast, the outcomes of the two experiments involved in tossing a coin twice are **independent**, and so the probability of getting a head on the second toss remains unchanged at 0.5, whatever the outcome of the first toss.

Although you may find it helpful to think about combined events in terms of how they would be represented on a tree diagram, you may not always actually draw them in this way. If there are several experiments and perhaps more than two possible outcomes from each, drawing a tree diagram can be very time-consuming.

Example 3.8

Free our David online campaign

Is this justice?

In 2012, David Starr was sentenced to 12 years' imprisonment for armed robbery solely on the basis of an identification parade. He was one of 12 people in the parade and was picked out by one witness but not by three others.

Many people who knew David well believe he was incapable of such a crime. Please add your voice to the clamour for a review of his case by clicking on the 'Free David' button.

Free David

How conclusive is this sort of evidence, or, to put it another way, how likely is it that a mistake has been made?

Investigate the likelihood that David Starr really did commit the robbery.

Solution

In this situation you need to assess the probability of an innocent individual being picked out by chance alone. Assume that David Starr was innocent and the witnesses were selecting in a purely random way (that is, with a probability of $\frac{1}{12}$ of selecting each person and a probability of $\frac{11}{12}$ of not selecting each person). If each of the witnesses selected just one of the twelve people in the identity parade in this random manner, how likely is it that David Starr would be picked out by at least one witness?

$$P(\text{at least one selection}) = 1 - P(\text{no selections})$$
$$= 1 - \frac{11}{12} \times \frac{11}{12} \times \frac{11}{12} \times \frac{11}{12}$$
$$= 1 - 0.706 = 0.294 \text{ (i.e. roughly 30\%).}$$

In other words, there is about a 30% chance of an innocent person being chosen in this way by at least one of the witnesses.

The website concluded:

> Is 30% really the sort of figure we have in mind when judges use the phrase 'beyond reasonable doubt'? Because if it is, many innocent people will be condemned to a life behind bars.

This raises an important statistical idea, which you will meet again if you study *Probability & Statistics 2*, about how we make judgements and decisions.

Judgements are usually made under conditions of uncertainty and involve us in having to weigh up the plausibility of one explanation against that of another. Statistical judgements are usually made on such a basis. We choose one explanation if we judge the alternative explanation to be sufficiently unlikely, that is if the probability of its being true is sufficiently small. Exactly how small this probability has to be will depend on the individual circumstances and is called the **significance level**.

Exercise 3B

1   The probability of a pregnant woman giving birth to a girl is about 0.49.

Draw a tree diagram showing the possible outcomes if she has two babies (not twins).

From the tree diagram, calculate the following probabilities:

(i)   that the babies are both girls

(ii) that the babies are the same sex

(iii) that the second baby is of different sex from the first.

2   In a certain district of a large city, the probability of a household suffering a break-in in a particular year is 0.07 and the probability of its car being stolen is 0.12.

Assuming these two trials are independent of each other, draw a tree diagram showing the possible outcomes for a particular year.

Calculate, for a randomly selected household with one car, the following probabilities:

(i)   that the household is a victim of both crimes during that year

(ii) that the household suffers *only one* of these misfortunes during that year

(iii) that the household suffers *at least one* of these misfortunes during that year.

3 There are 12 people at an identification parade. Three witnesses are called to identify the accused person.

Assuming they make their choice purely by random selection, draw a tree diagram showing the possible events.

(i) From the tree diagram, calculate the following probabilities:

(a) that all three witnesses select the accused person

(b) that none of the witnesses selects the accused person

(c) that at least two of the witnesses select the accused person.

(ii) Suppose now that by changing the composition of people in the identification parade, the first two witnesses increase their chances of selecting the accused person to 0.25.

Draw a new tree diagram and calculate the following probabilities:

(a) that all three witnesses select the accused person

(b) that none of the witnesses selects the accused person

(c) that at least two of the witnesses select the accused person.

4 Ruth drives her car to work – provided she can get it to start! When she remembers to put the car in the garage the night before, it starts next morning with a probability of 0.95. When she forgets to put the car away, it starts next morning with a probability of 0.75. She remembers to garage her car 90% of the time.

What is the probability that Ruth drives her car to work on a randomly chosen day?

 5 Around 0.8% of men are red–green colour-blind (the figure is slightly different for women) and roughly 1 in 5 men is left-handed.

Assuming these characteristics are inherited independently, calculate with the aid of a tree diagram the probability that a man chosen at random will:

(i) be both colour-blind and left-handed

(ii) be colour-blind and not left-handed

(iii) be colour-blind or left-handed

(iv) be neither colour-blind nor left-handed.

6 Three dice are thrown. Find the probability of obtaining:

(i) at least two 6s

(ii) no 6s

(iii) different scores on all the dice.

 7 Explain the flaw in this argument and rewrite it as a valid statement.

The probability of throwing a 6 on a fair die $= \frac{1}{6}$. Therefore the probability of throwing at least one 6 in six throws of the die is $\frac{1}{6} + \frac{1}{6} + \frac{1}{6} + \frac{1}{6} + \frac{1}{6} + \frac{1}{6} = 1$ so it is a certainty.

8 Two dice are thrown. The scores on the dice are added.

(i) Copy and complete this sample space diagram.

		First die					
		1	2	3	4	5	6
Second die	1						
	2						
	3						
	4						10
	5						11
	6	7	8	9	10	11	12

(ii) What is the probability of a score of 4?

(iii) What is the most likely outcome?

(iv) Criticise this argument:

There are 11 possible outcomes, 2, 3, 4, up to 12. Therefore each of them has a probability of $\frac{1}{11}$.

9 The probability of someone catching flu in a particular winter when they have been given the flu vaccine is 0.1. Without the vaccine, the probability of catching flu is 0.4. If 30% of the population has been given the vaccine, what is the probability that a person chosen at random from the population will catch flu over that winter?

3.6 Conditional probability

Sad news

Myra

My best friend had a heart attack while out shopping. Sachit was rushed to hospital but died on the way. He was only 47 – too young ☹

What is the probability that somebody chosen at random will die of a heart attack in the next 12 months?

One approach would be to say that, since there are about 300 000 deaths per year from heart and circulatory diseases (H & CD) among the 57 000 000 population of the country where Sachit lived,

$$\text{probability} = \frac{\text{number of deaths from H \& CD per year}}{\text{total population}}$$

$$= \frac{300\,000}{57\,000\,000} = 0.0053.$$

However, if you think about it, you will probably realise that this is rather a meaningless figure. For a start, young people are much less at risk than those in or beyond middle age.

3

So you might wish to give two answers:

$$P_1 = \frac{\text{deaths from H \& CD among over–40s}}{\text{population of over–40s}}$$

$$P_2 = \frac{\text{deaths from H \& CD among under–40s}}{\text{population of under–40s}}$$

Typically only 1500 of the deaths would be among the under–40s, leaving (on the basis of these figures) 298 500 among the over–40s. About 25 000 000 people in the country are over 40, and 32 000 000 under 40 (40 years and 1 day counts as over 40). This gives

$$P_1 = \frac{\text{deaths from H \& CD among over–40s}}{\text{population of over–40s}}$$

$$= \frac{298\,500}{25\,000\,000}$$

$$= 0.0119$$

and $\quad P_2 = \dfrac{\text{deaths from H \& CD among under–40s}}{\text{population of under–40s}}$

$$= \frac{1500}{32\,000\,000}$$

$$= 0.000\,047.$$

So somebody in the older group is over 200 times more likely to die of a heart attack than somebody in the younger group. Putting them both together as an average figure resulted in a figure that was representative of neither group.

But why stop there? You could, if you had the figures, divide the population up into 10-year, 5-year, or even 1-year intervals. That would certainly improve the accuracy; but there are also more factors that you might wish to take into account, such as the following.

▸▸ Is the person overweight?

▸▸ Does the person smoke?

▸▸ Does the person take regular exercise?

The more conditions you build in, the more accurate the estimate of the probability.

You can see how the conditions are brought in by looking at P_1:

$$P_1 = \frac{\text{deaths from H \& CD among over–40s}}{\text{population of over–40s}}$$

$$= \frac{298\,500}{25\,000\,000}$$

$$= 0.0119$$

You would write this in symbols as follows:

Event G: Somebody selected at random is over 40.
Event H: Somebody selected at random dies from H & CD.

The probability of someone dying from H & CD given that he or she is over 40 is given by the conditional probability $P(H \mid G)$.

$$P(H|G) = \frac{n(H \cap G)}{n(G)}$$

$P(H \mid G)$ means the probability of event H occurring *given that* event G has occurred.

$$= \frac{n(H \cap G) / n(\mathcal{E})}{n(G) / n(\mathcal{E})}$$

$$= \frac{P(H \cap G)}{P(G)}.$$

This result may be written in general form for all cases of conditional probability for events A and B.

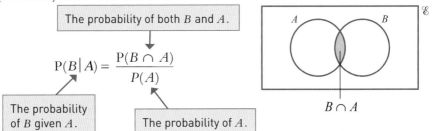

The probability of both B and A.

$$P(B|A) = \frac{P(B \cap A)}{P(A)}$$

The probability of B given A.

The probability of A.

$B \cap A$

▲ Figure 3.7

Conditional probability is used when your estimate of the probability of an event is altered by your knowledge of whether some other event has occurred. In this case the estimate of the probability of somebody dying from heart and circulatory diseases, $P(H)$, is altered by a knowledge of whether or not the person is over 40.

Thus conditional probability addresses the question of whether one event is dependent on another one. If the probability of event B is not affected by the occurrence of event A, we say that B is **independent** of A. If, on the other hand, the probability of event B is affected by the occurrence (or not) of event A, we say that B is **dependent** on A.

» If A and B are independent, then $P(B|A) = P(B|A')$ and this is just $P(B)$.

» If A and B are dependent, then $P(B|A) \neq P(B|A')$.

As you have already seen, the probability of a combined event is the product of the separate probabilities of each event, provided the question of dependence between the two events is properly dealt with. Specifically:

The probability of both A and B occurring.

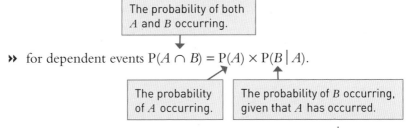

» for dependent events $P(A \cap B) = P(A) \times P(B|A)$.

The probability of A occurring.

The probability of B occurring, given that A has occurred.

When A and B are independent events, then, because $P(B|A) = P(B)$, this can be written as:

» for independent events $P(A \cap B) = P(A) \times P(B)$.

Example 3.9

A company is worried about the high turnover of its employees and decides to investigate whether they are more likely to stay if they are given training. On 1 January one year the company was employing 256 people (excluding those about to retire). During that year a record was kept of who received training as well as who left the company. The results are summarised in this table.

	Still employed	Left company	Total
Given training	109	43	152
Not given training	60	44	104
Totals	169	87	256

(i) Find the probability that a randomly selected employee:

    (a) received training

    (b) received training and did not leave the company.

(ii) Are the events T and S independent?

(iii) Find the probability that a randomly selected employee:

    (a) did not leave the company, given that the person had received training

    (b) did not leave the company, given that the person had not received training.

Solution

Using the notation T: The employee received training

                       S: The employee stayed in the company

(i) (a) $P(T) = \dfrac{n(T)}{n(\mathscr{E})} = \dfrac{152}{256} = 0.594$

    (b) $P(T \cap S) = \dfrac{n(T \cap S)}{n(\mathscr{E})} = \dfrac{109}{256} = 0.426$

(ii) If T and S are independent events then $P(T \cap S) = P(T) \times P(S)$.

$P(S) = \dfrac{n(S)}{n(\mathscr{E})} = \dfrac{169}{256} = 0.660$

$P(T) \times P(S) = \dfrac{152}{256} \times \dfrac{169}{256} = 0.392$

As $P(T \cap S) \neq P(T) \times P(S)$, the events T and S are not independent.

(iii) (a) $P(S|T) = \dfrac{P(S \cap T)}{P(T)} = \dfrac{\frac{109}{256}}{\frac{152}{256}} = \dfrac{109}{152} = 0.717$

    (b) $P(S|T') = \dfrac{P(S \cap T')}{P(T')} = \dfrac{\frac{60}{256}}{\frac{104}{256}} = \dfrac{60}{104} = 0.577$

Since $P(S \mid T)$ is not the same as $P(S \mid T')$, the event S is not independent of the event T. Each of S and T is dependent on the other, a conclusion which matches common sense. It is almost certainly true that training increases employees' job satisfaction and so makes them more likely to stay, but it is also probably true that the company is more likely to go to the expense of training the employees who seem less inclined to move on to other jobs.

?

> How would you show that the event T is not independent of the event S?

In some situations you may find it helps to represent a problem such as this as a Venn diagram.

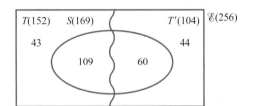

$T(152)$ $S(169)$ $T'(104)$ $\mathscr{E}(256)$
43
109   60
44

▲ Figure 3.8

?

> What do the various numbers and letters represent?
> Where is the region S'?
> How are the numbers on the diagram related to the answers to parts (i) to (iii)?

In other situations it may be helpful to think of conditional probabilities in terms of tree diagrams. Conditional probabilities are needed when events are *dependent*, that is when the outcome of one trial affects the outcomes from a subsequent trial, so, for dependent events, the probabilities of all but the first layer of a tree diagram will be conditional.

Example 3.10

Rebecca is buying two goldfish from a pet shop. The shop's tank contains seven male fish and eight female fish but they all look the same.

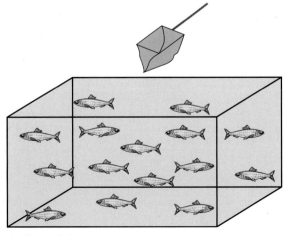

▲ Figure 3.9

Find the probability that Rebecca's fish are:

(i) both the same sex

(ii) both female

(iii) both female given that they are the same sex.

Solution

The situation is shown on this tree diagram.

first fish   second fish

$\frac{7}{15}$ M

$\frac{6}{14}$ M   P(both male) $= \frac{7}{15} \times \frac{6}{14} = \frac{42}{210}$

$\frac{8}{14}$ F   P(male, female) $= \frac{7}{15} \times \frac{8}{14} = \frac{56}{210}$

$\frac{8}{15}$ F

$\frac{7}{14}$ M   P(female, male) $= \frac{8}{15} \times \frac{7}{14} = \frac{56}{210}$

$\frac{7}{14}$ F   P(both female) $= \frac{8}{15} \times \frac{7}{14} = \frac{56}{210}$

▲ **Figure 3.10**

(i) P(both the same sex) = P(both male) + P(both female)

$$= \frac{42}{210} + \frac{56}{210} = \frac{98}{210} = \frac{7}{15}$$

(ii) P(both female) $= \frac{56}{210} = \frac{4}{15}$

(iii) P(both female | both the same sex)

$$= \text{P(both female and the same sex)} \div \text{P(both the same sex)} = \frac{\frac{4}{15}}{\frac{7}{15}} = \frac{4}{7}$$

This is the same as P(both female).

The ideas in the last example can be expressed more generally for any two dependent events, A and B. The tree diagram would be as shown in Figure 3.11.

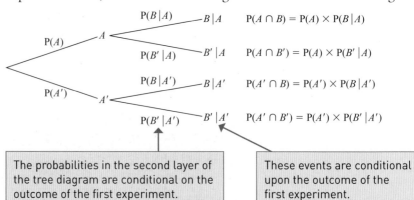

P(A)   A

P(B|A)   B|A $P(A \cap B) = P(A) \times P(B|A)$

P(B'|A)   B'|A $P(A \cap B') = P(A) \times P(B'|A)$

P(A')   A'

P(B|A')   B|A' $P(A' \cap B) = P(A') \times P(B|A')$

P(B'|A')   B'|A' $P(A' \cap B') = P(A') \times P(B'|A')$

The probabilities in the second layer of the tree diagram are conditional on the outcome of the first experiment.

These events are conditional upon the outcome of the first experiment.

▲ **Figure 3.11**

The tree diagram shows you that:

» $P(B) = P(A \cap B) + P(A' \cap B)$
$= P(A) \times P(B \mid A) + P(A') \times P(B \mid A')$

» $P(A \cap B) = P(A) \times P(B \mid A)$

$\Rightarrow \quad P(B \mid A) = \dfrac{P(A \cap B)}{P(A)}$

> How were these results used in Example 3.10 about the goldfish?

Exercise 3C

1 In a school of 600 students, 360 are girls. There are 320 hockey players, of whom 200 are girls. Among the hockey players there are 28 goalkeepers, 19 of them girls. Find the probability that:

(i)   a student chosen at random is a girl

(ii)   a girl chosen at random plays hockey

(iii) a hockey player chosen at random is a girl

(iv)   a student chosen at random is a goalkeeper

(v)   a goalkeeper chosen at random is a boy

(vi)   a male hockey player chosen at random is a goalkeeper

(vii) a hockey player chosen at random is a male goalkeeper

(viii) two students chosen at random are both goalkeepers

(ix)   two students chosen at random are a male goalkeeper and a female goalkeeper

(x)   two students chosen at random are one boy and one girl.

2 100 cars are entered for a road-worthiness test which is in two parts, mechanical and electrical. A car passes only if it passes both parts. Half the cars fail the electrical test and 62 pass the mechanical. 15 pass the electrical but fail the mechanical test.

Find the probability that a car chosen at random:

(i)   passes overall

(ii)   fails on one test only

(iii) given that it has failed, failed the mechanical test only.

3 Two dice are thrown. What is the probability that the total is:

(i)   7

(ii)   a prime number

(iii) 7, given that it is a prime number?

CP

4 A and B are two events with probabilities given by $P(A) = 0.4$, $P(B) = 0.7$ and $P(A \cap B) = 0.35$.

(i)   Find $P(A \mid B)$ and $P(B \mid A)$.

(ii)   Show that the events A and B are not independent.

PS **5** Quark hunting is a dangerous occupation. On a quark hunt, there is a probability of $\frac{1}{4}$ that the hunter is killed. The quark is twice as likely to be killed as the hunter. There is a probability of $\frac{1}{3}$ that both survive.

(i) Copy and complete this table of probabilities.

	Hunter dies	Hunter lives	Total
Quark dies			$\frac{1}{2}$
Quark lives		$\frac{1}{3}$	$\frac{1}{2}$
Total	$\frac{1}{4}$		1

Find the probability that:

(ii) both the hunter and the quark die

(iii) the hunter lives and the quark dies

(iv) the hunter lives, given that the quark dies.

6 There are 90 players in a tennis club. Of these, 23 are juniors, the rest are seniors. 34 of the seniors and 10 of the juniors are male. There are 8 juniors who are left-handed, 5 of whom are male. There are 18 left-handed players in total, 4 of whom are female seniors.

(i) Represent this information in a Venn diagram.

(ii) What is the probability that:

(a) a male player selected at random is left-handed?

(b) a left-handed player selected at random is a female junior?

(c) a player selected at random is either a junior or a female?

(d) a player selected at random is right-handed?

(e) a right-handed player selected at random is not a junior?

(f) a right-handed female player selected at random is a junior?

7 Data about employment for males and females in a small rural area are shown in the table.

	Unemployed	Employed
Male	206	412
Female	358	305

A person from this area is chosen at random. Let M be the event that the person is male and let E be the event that the person is employed.

(i) Find P(M).

(ii) Find P(M and E).

(iii) Are M and E independent events? Justify your answer.

(iv) Given that the person chosen is unemployed, find the probability that the person is female.

Cambridge International AS & A Level Mathematics
9709 Paper 6 Q5 June 2005

8 The probability that Henk goes swimming on any day is 0.2. On a day when he goes swimming, the probability that Henk has burgers for supper is 0.75. On a day when he does not go swimming, the probability that he has burgers for supper is x. This information is shown on the following tree diagram.

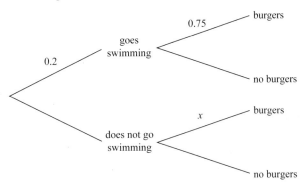

The probability that Henk has burgers for supper on any day is 0.5.

(i) Find x.

(ii) Given that Henk has burgers for supper, find the probability that he went swimming that day.

Cambridge International AS & A Level Mathematics
9709 Paper 6 Q2 June 2006

9 Boxes of sweets contain toffees and chocolate. Box A contains 6 toffees and 4 chocolates, box B contains 5 toffees and 3 chocolates, and box C contains 3 toffees and 7 chocolates. One of the boxes is chosen at random and two sweets are taken out, one after the other, and eaten.

(i) Find the probability that they are both toffees.

(ii) Given that they are both toffees, find the probability that they both come from box A.

Cambridge International AS & A Level Mathematics
9709 Paper 6 Q2 November 2005

10 There are three sets of traffic lights on Karinne's journey to work. The independent probabilities that Karinne has to stop at the first, second and third set of lights are 0.4, 0.8 and 0.3 respectively.

(i) Draw a tree diagram to show this information.

(ii) Find the probability that Karinne has to stop at each of the first two sets of lights but does not have to stop at the third set.

(iii) Find the probability that Karinne has to stop at exactly two of the three sets of lights.

(iv) Find the probability that Karinne has to stop at the first set of lights, given that she has to stop at exactly two sets of lights.

Cambridge International AS & A Level Mathematics
9709 Paper 6 Q6 November 2008

11 A survey is undertaken to investigate how many photos people take on a one-week holiday and also how many times they view past photos. For a randomly chosen person, the probability of taking fewer than 100 photos is x. The probability that these people view past photos at least 3 times is 0.76.

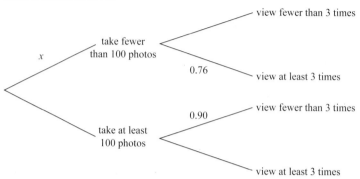

For those who take at least 100 photos, the probability that they view past photos fewer than 3 times is 0.90. This information is shown in the tree diagram. The probability that a randomly chosen person views past photos fewer than 3 times is 0.801.

(i) Find x.

(ii) Given that a person views past photos at least 3 times, find the probability that this person takes at least 100 photos.

Cambridge International AS & A Level Mathematics
9709 Paper 61 Q4 June 2015

12 Playground equipment consists of swings (S), roundabouts (R), climbing frames (C) and play-houses (P). The numbers of pieces of equipment in each of 3 playgrounds are as follows.

Playground X	Playground Y	Playground Z
$3S, 2R, 4P$	$6S, 3R, 1C, 2P$	$8S, 3R, 4C, 1P$

Each day Nur takes her child to one of the playgrounds. The probability that she chooses playground X is $\frac{1}{4}$. The probability that she chooses playground Y is $\frac{1}{4}$. The probability that she chooses playground Z is $\frac{1}{2}$. When she arrives at the playground, she chooses one piece of equipment at random.

(i) Find the probability that Nur chooses a play-house.

(ii) Given that Nur chooses a climbing frame, find the probability that she chose playground Y.

Cambridge International AS & A Level Mathematics
9709 Paper 61 Q5 June 2014

Estimating minnows

A building company is proposing to fill in a pond that is the home for many different species of wildlife. The local council commission a naturalist called Henry to do a survey of the wildlife that is dependent on the pond. As part of this exercise Henry decides to estimate how many minnows are living in the pond. The minnows cannot all be seen at the same time so it is not possible just to count them.

This task involves the various stages of the problem solving cycle.

1 **Problem specification and analysis**

Henry starts by planning how he will go about the task.

The procedure he chooses is based on a method called **capture–recapture**. To carry it out he needs a minnow trap and some suitable markers for minnows.

2 **Information collection**

Henry sets the trap once a day for a week, starting on Sunday. Each time that he opens the trap he counts how many minnows he caught and how many of them are already marked. He then marks those that are not already marked and returns the minnows to the pond.

The following table shows his results for the week.

Day	Sun	Mon	Tues	Wed	Thurs	Fri	Sat
Number caught	10	12	8	15	6	12	16
Number already marked	–	1	1	3	2	4	6

3 **Processing and representation**

   (i) After Henry has returned the minnows to the pond on Monday, he estimates that there are 120 minnows in the pond. After Tuesday's catch he makes a new estimate of 168 minnows. Show how he calculates these figures.

   (ii) Use the figures for the subsequent days' catches to make four more estimates of the number of minnows in the pond.

   (iii) Draw a suitable diagram to illustrate the estimates.

4 **Interpretation**

   (i) Henry has to write a report to the council. It will include a short section about the minnows. Comment briefly on what it might say about the following:

      (a) The best estimate of the number of minnows

      (b) Any assumptions required for the calculations and if they are reasonable

      (c) How accurate the estimate is likely to be.

   (ii) Suggest a possible improvement to Henry's method for data collection.

KEY POINTS

1   The probability of an event A *is*

$$P(A) = \frac{n(A)}{n(\mathscr{E})}$$

where $n(A)$ is the number of ways that A can occur and $n(\mathscr{E})$ is the total number of ways that all possible events can occur, all of which are equally likely.

2   A sample space is a list of all possible outcomes of an experiment.

3   For any two events, A and B, of the same experiment,

$$P(A \cup B) = P(A) + P(B) - P(A \cap B).$$

Where the events are *mutually exclusive* (i.e. where the events do not overlap) the rule still holds but, since $P(A \cap B)$ is now equal to zero, the equation simplifies to:

$$P(A \cup B) = P(A) + P(B).$$

4   Where an experiment produces two or more mutually exclusive events, the probabilities of the separate events sum to 1.

5   $P(A) + P(A') = 1$

6   For two independent events, A and B,

$$P(A \cap B) = P(A) \times P(B).$$

7   $P(B \mid A)$ means the probability of event B occurring given that event A has already occurred,

$$P(B \mid A) = \frac{P(A \cap B)}{P(A)}.$$

8   The probability that event A and then event B occur, in that order, is $P(A) \times P(B \mid A)$.

9   If event B is independent of event A,

$$P(B \mid A) = P(B \mid A') = P(B).$$

LEARNING OUTCOMES

Now that you have finished this chapter, you should be able to

- measure probability using the number of ways an event can happen and the number of equally likely outcomes

- interpret probabilities of 0 and 1

- complete a sample space diagram to list all possible outcomes

- find the probability of the complement of an event

- use Venn diagrams in probability problems

- understand the terms mutually exclusive and independent

- determine whether a pair of events are mutually exclusive or independent

- use addition and multiplication of probabilities

- understand that $P(A \cup B) = P(A) + P(B) - P(A \cap B)$

- solve probability problems with two events

- use a tree diagram to solve probability problems with two events

- solve problems involving conditional probability

- use the formula $P(A \mid B) = \dfrac{P(A \cap B)}{P(B)}$.

4

Discrete random variables

> An approximate answer to the right problem is worth a good deal more than an exact answer to an approximate problem.
> *John Tukey (1915–2000)*

Car Share World

Share life's journey!

Towns and cities around the country are gridlocked with traffic – many of these cars have just one occupant. To solve this problem, Car Share World is launching a new scheme for people to carshare on journeys into all major cities. Our comprehensive database can put interested drivers in touch with each other and live updates via your mobile will display the number of car shares available in any major city. Car shares are available from centralised locations for maximum convenience.

Car Share World is running a small trial scheme in a busy town just south of the capital. We will be conducting a survey to measure the success of the trial. Keep up to date with the trial via the latest news on our website.

> ❯ How would you collect information on the volume of traffic in the town?

A traffic survey, at critical points around the town centre, was conducted at peak travelling times over a period of a working week. The survey involved 1000 cars. The number of people in each car was noted, with the following results.

Number of people per car	1	2	3	4	5	> 5
Frequency	560	240	150	40	10	0

> ❯ How would you illustrate such a distribution?
> ❯ What are the main features of this distribution?

The numbers of people per car are necessarily discrete. A discrete frequency distribution is best illustrated by a vertical line chart, as in Figure 4.1. This shows you that the distribution has positive skew, with most of the data at the lower end of the distribution.

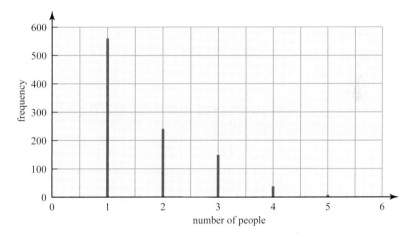

▲ **Figure 4.1**

The survey involved 1000 cars. This is a large sample and so it is reasonable to use the results to estimate the **probabilities** of the various possible outcomes: 1, 2, 3, 4, 5 people per car. You divide each frequency by 1000 to obtain the **relative frequency**, or probability, of each outcome (number of people).

Outcome (Number of people)	1	2	3	4	5	> 5
Probability (Relative frequency)	0.56	0.24	0.15	0.04	0.01	0

4.1 Discrete random variables

You now have a **mathematical model** to describe a particular situation. In statistics you are often looking for models to describe and explain the data you find in the real world. In this chapter you are introduced to some of the techniques for working with models for discrete data. Such models use **discrete random variables**.

The model is **discrete** since the number of passengers can be counted and takes positive integer values only. The number of passengers is a **random variable** since the actual value of the outcome is variable and can only be predicted with a given probability, i.e. the outcomes occur at random.

Discrete random variables may have a **finite** or an **infinite** number of possible outcomes.

The distribution we have outlined so far is finite – in the survey the maximum number of people observed was five, but the maximum could be, say, eight, depending on the size of car. In this case there would be eight possible outcomes. A well-known example of a finite discrete random variable occurs in the **binomial distribution**, which you will study in Chapter 6.

On the other hand, if you considered the number of hits on a website in a given day, there may be no theoretical maximum, in which case the distribution may be considered as infinite. A well-known example of an infinite discrete random variable occurs in the **Poisson distribution**, which you will meet if you study *Probability & Statistics 2*.

The study of discrete random variables in this chapter will be limited to finite cases.

Notation and conditions for a discrete random variable

A discrete random variable is usually denoted by an upper case letter, such as X, Y or Z. You may think of this as the name of the variable. The particular values that the variable takes are denoted by lower case letters, such as r. Sometimes these are given suffixes $r_1, r_2, r_3, \ldots$. Thus $P(X = r_1)$ is the probability that the discrete random variable X takes the particular value r_1. The expression $P(X = r)$ is used to express a more general idea, as, for example, in a table heading.

Another, shorter way of writing probabilities is $p_1, p_2, p_3, \ldots$. If a finite discrete random variable has n distinct outcomes $r_1, r_2, \ldots, r_n$, with associated probabilities $p_1, p_2, \ldots, p_n$, then the sum of the probabilities must equal 1. Since the various outcomes cover all possibilities, they are **exhaustive**.

Formally we have:

$$p_1 + p_2 + \ldots + p_n = 1$$

$$\text{or } \sum_{i=1}^{n} p_i = \sum_{i=1}^{n} P(X = r_i) = 1.$$

> You should be familiar with all these notations.

If there is no ambiguity then $\sum_{i=1}^{n} P(X = r_i)$ is often abbreviated to $\sum P(X = r)$ or p_r.

You will often see an alternative notation used, in which the values that the variable takes are denoted by x rather than r. In this book, r is used for a discrete variable and in *Probability & Statistics 2*, x is used for a continuous variable.

Diagrams of discrete random variables

Just as with frequency distributions for discrete data, the most appropriate diagram to illustrate a discrete random variable is a vertical line chart. Figure 4.2 shows a diagram of the probability distribution of X, the number of people per car. Note that it is identical in shape to the corresponding frequency diagram in Figure 4.1. The only real difference is the change of scale on the vertical axis.

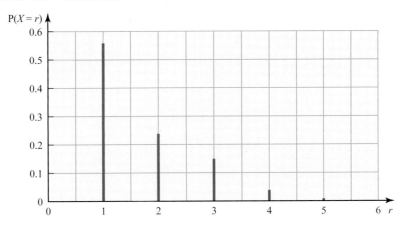

▲ **Figure 4.2**

| Example 4.1 | Two tetrahedral dice, each with faces labelled 1, 2, 3 and 4, are thrown and the random variable X represents the sum of the numbers shown on the dice. |

(i)   Find the probability distribution of X.
(ii) Illustrate the distribution and describe the shape of the distribution.
(iii) What is the probability that any throw of the dice results in a value of X which is an odd number?

Solution

(i)   The table shows all the possible totals when the two dice are thrown.

		First die			
		1	2	3	4
Second die	1	2	3	4	5
	2	3	4	5	6
	3	4	5	6	7
	4	5	6	7	8

You can use the table to write down the probability distribution for X.

r	2	3	4	5	6	7	8
$P(X = r)$	$\frac{1}{16}$	$\frac{2}{16}$	$\frac{3}{16}$	$\frac{4}{16}$	$\frac{3}{16}$	$\frac{2}{16}$	$\frac{1}{16}$

➡

(ii) The vertical line chart in Figure 4.3 illustrates this distribution, which is symmetrical.

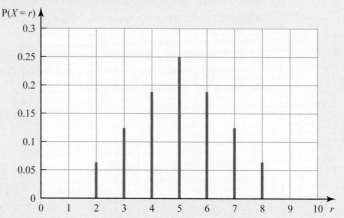

▲ Figure 4.3

(iii) The probability that X is an odd number

$$= P(X = 3) + P(X = 5) + P(X = 7)$$
$$= \frac{2}{16} + \frac{4}{16} + \frac{2}{16}$$
$$= \frac{1}{2}$$

As well as defining a discrete random variable by tabulating the probability distribution, another effective way is to use an algebraic definition of the form $P(X = r) = f(r)$ for given values of r.

The following example illustrates how this may be used.

Example 4.2

The probability distribution of a random variable X is given by

$$P(X = r) = kr \qquad \text{for } r = 1, 2, 3, 4$$
$$P(X = r) = 0 \qquad \text{otherwise.}$$

(i) Find the value of the constant k.
(ii) Illustrate the distribution and describe the shape of the distribution.
(iii) Two successive values of X are generated independently of each other. Find the probability that:
  (a) both values of X are the same
  (b) the total of the two values of X is greater than 6.

Solution

(i) Tabulating the probability distribution for X gives:

r	1	2	3	4
$P(X = r)$	k	$2k$	$3k$	$4k$

$$\sum P(X=r) \quad\quad = 1$$
$$\Rightarrow k + 2k + 3k + 4k = 1$$
$$\Rightarrow \quad\quad\quad\quad 10k = 1$$
$$\Rightarrow \quad\quad\quad\quad\quad k = 0.1$$

Hence $P(X=r) = 0.1r$, for $r = 1, 2, 3, 4$, which gives the following probability distribution.

r	1	2	3	4
P(X = r)	0.1	0.2	0.3	0.4

(ii) The vertical line chart in Figure 4.4 illustrates this distribution. It has negative skew.

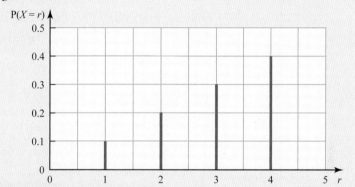

▲ **Figure 4.4**

(iii) Let X_1 represent the first value generated and X_2 the second value generated.

(a) P(both values of X are the same)

$$= P(X_1 = X_2 = 1 \text{ or } X_1 = X_2 = 2 \text{ or } X_1 = X_2 = 3 \text{ or } X_1 = X_2 = 4)$$
$$= P(X_1 = X_2 = 1) + P(X_1 = X_2 = 2) + P(X_1 = X_2 = 3)$$
$$\quad + P(X_1 = X_2 = 4)$$
$$= P(X_1 = 1) \times P(X_2 = 1) + P(X_1 = 2) \times P(X_2 = 2)$$
$$\quad + P(X_1 = 3) \times P(X_2 = 3) + P(X_1 = 4) \times P(X_2 = 4)$$
$$= (0.1)^2 + (0.2)^2 + (0.3)^2 + (0.4)^2$$
$$= 0.01 + 0.04 + 0.09 + 0.16$$
$$= 0.3$$

(b) P(total of the two values is greater than 6)

$$= P(X_1 + X_2 > 6)$$
$$= P(X_1 + X_2 = 7 \text{ or } 8)$$
$$= P(X_1 + X_2 = 7) + P(X_1 + X_2 = 8)$$
$$= P(X_1 = 3) \times P(X_2 = 4) + P(X_1 = 4) \times P(X_2 = 3)$$
$$\quad + P(X_1 = 4) \times P(X_2 = 4)$$
$$= 0.3 \times 0.4 + 0.4 \times 0.3 + 0.4 \times 0.4$$
$$= 0.12 + 0.12 + 0.16$$
$$= 0.4$$

1   The random variable X is given by the sum of the scores when two ordinary dice are thrown.

(i)   Find the probability distribution of X.

(ii)   Illustrate the distribution and describe the shape of the distribution.

(iii) Find the values of:

(a) $P(X > 8)$

(b) $P(X$ is even)

(c) $P(|X - 7| < 3)$.

CP 2   The random variable Y is given by the absolute difference between the scores when two ordinary dice are thrown.

(i)   Find the probability distribution of Y.

(ii)   Illustrate the distribution and describe the shape of the distribution.

(iii) Find the values of:

(a) $P(Y < 3)$

(b) $P(Y$ is odd).

3   The probability distribution of a discrete random variable X is given by:

$$P(X = r) = \frac{kr}{8} \qquad \text{for } r = 2, 4, 6, 8$$
$$P(X = r) = 0 \qquad \text{otherwise.}$$

(i)   Find the value of k and tabulate the probability distribution.

(ii)   If two successive values of X are generated independently find the probability that:

(a)   the two values are equal

(b)   the first value is greater than the second value.

CP 4   An irregular die with six faces produces scores, X, for which the probability distribution is given by:

$$P(X = r) = \frac{k}{r} \qquad \text{for } r = 1, 2, 3, 4, 5, 6$$
$$P(X = r) = 0 \qquad \text{otherwise.}$$

(i)   Find the value of k and illustrate the distribution.

(ii)   Show that, when this die is thrown twice, the probability of obtaining two equal scores is very nearly $\frac{1}{4}$.

CP 5   Three fair coins are tossed.

(i)   By considering the set of possible outcomes, HHH, HHT, etc., tabulate the probability distribution for X, the number of heads occurring.

(ii)   Illustrate the distribution and describe the shape of the distribution.

(iii) Find the probability that there are more heads than tails.

(iv)   Without further calculation, state whether your answer to part (iii) would be the same if four fair coins were tossed. Give a reason for your answer.

6 Two fair tetrahedral dice, each with faces labelled 1, 2, 3 and 4, are thrown and the random variable X is the product of the numbers shown on the dice.

(i) Find the probability distribution of X.

(ii) What is the probability that any throw of the dice results in a value of X which is an odd number?

7 An ornithologist carries out a study of the number of eggs laid per pair by a species of rare bird in its annual breeding season. He concludes that it may be considered as a discrete random variable X with probability distribution given by

$$P(X = 0) = 0.2$$
$$P(X = r) = k(4r - r^2) \qquad \text{for } r = 1, 2, 3, 4$$
$$P(X = r) = 0 \qquad \text{otherwise.}$$

(i) Find the value of k and write the probability distribution as a table.

The ornithologist observes that the probability of survival (that is of an egg hatching and of the chick living to the stage of leaving the nest) is dependent on the number of eggs in the nest. He estimates the probabilities to be as follows.

r	Probability of survival
1	0.8
2	0.6
3	0.4

(ii) Find, in the form of a table, the probability distribution of the number of chicks surviving per pair of adults.

8 A sociologist is investigating the changing pattern of the number of children which women have in a country. She denotes the present number by the random variable X which she finds to have the following probability distribution.

r	0	1	2	3	4	5 +
$P(X = r)$	0.09	0.22	a	0.19	0.08	negligible

(i) Find the value of a.

She is keen to find an algebraic expression for the probability distribution and suggests the following model.

$$P(X = r) = k(r + 1)(5 - r) \qquad \text{for } r = 0, 1, 2, 3, 4, 5$$
$$P(X = r) = 0 \qquad \text{otherwise.}$$

(ii) Find the value of k for this model.

(iii) Compare the algebraic model with the probabilities she found, illustrating both distributions on one diagram.

Do you think it is a good model?

9 In a game, each player throws three ordinary six-sided dice. The random variable X is the largest number showing on the dice, so for example, for scores of 2, 5 and 4, $X = 5$.

(i) Find the probability that $X = 1$, i.e. $\mathrm{P}(X = 1)$.

(ii) Find $\mathrm{P}(X \leqslant 2)$ and deduce that $\mathrm{P}(X = 2) = \frac{7}{216}$.

(iii) Find $\mathrm{P}(X \leqslant r)$ and so deduce $\mathrm{P}(X = r)$, for $r = 3, 4, 5, 6$.

(iv) Illustrate and describe the probability distribution of X.

10 A box contains six black pens and four red pens. Three pens are taken at random from the box.

(i) By considering the selection of pens as sampling without replacement, illustrate the various outcomes on a probability tree diagram.

(ii) The random variable X represents the number of red pens obtained. Find the probability distribution of X.

11 A vegetable basket contains 12 peppers, of which 3 are red, 4 are green and 5 are yellow. Three peppers are taken, at random and without replacement, from the basket.

(i) Find the probability that the three peppers are all different colours.

(ii) Show that the probability that exactly 2 of the peppers taken are green is $\frac{12}{55}$.

(iii) The number of **green** peppers taken is denoted by the discrete random variable X. Draw up a probability distribution table for X.

Cambridge International AS & A Level Mathematics
9709 Paper 6 Q7 June 2007

4.2 Expectation and variance

Car Share World

Share life's journey!

Latest update ...

Car-share trial a massive success. Traffic volume down and number of occupants per car up!

> ❯ What statistical evidence do you think Car Share World's claim is based on?

A second traffic survey, at critical points around the town centre, was conducted at peak travelling times over a period of a working week. This time the survey involved 800 cars. The number of people in each car is shown in the table.

Number of people per car	1	2	3	4	5	> 5
Frequency	280	300	164	52	4	0

> ❯ How would you compare the results in the two traffic surveys?

The survey involved 800 cars. This is a fairly large sample and so, once again, it is reasonable to use the results to estimate the probabilities of the various possible outcomes: 1, 2, 3, 4 and 5 people per car, as before.

Outcome (Number of people)	1	2	3	4	5	> 5
Probability (Relative frequency)	0.35	0.375	0.205	0.065	0.005	0

One way to compare the two probability distributions, before and after the car-sharing campaign, is to calculate a measure of central tendency and a measure of spread.

The most useful measure of central tendency is the **mean** or **expectation** of the random variable and the most useful measure of spread is the **variance**. To a large extent the calculation of these statistics mirrors the corresponding statistics for a frequency distribution, $\bar{x}$ and sd^2.

> ▶ **ACTIVITY 4.1**
> ....
>
> Find the mean and variance of the frequency distribution for the people-per-car survey following the introduction of the car-sharing scheme.

Using relative frequencies generates an alternative approach which gives the *expectation* $E(X) = \mu$ and *variance* $\text{Var}(X) = \sigma^2$ for a discrete random variable.

We define the expectation, $E(X)$ as

$$E(X) = \mu = \sum r P(X = r) = \sum r p_r$$

and variance, $\text{Var}(X)$ as

> Notice the notation, μ for the distribution's mean and σ for its standard deviation. Also notice the shortened notation for $P(X = r)$.

$$\sigma^2 = E([X - \mu]^2) = \sum (r - \mu)^2 \, p_r$$

or

$$\sigma^2 = E(X^2) - \mu^2 = \sum r^2 p_r - \left[\sum r p_r\right]^2.$$

> σ^2 is read as 'sigma squared'.

The second version of the variance is often written as $E(X^2) - [E(X)]^2$, which can be remembered as 'the expectation of the squares minus the square of the expectation'.

These formulae can also be written as:

$$E(X) = \sum xp$$

$$\text{Var}(X) = \sum x^2 p - [E(X)]^2$$

Look at how expectation and variance are calculated using the probability distribution developed from the second survey of number of people per car. You can use these statistics to compare the distribution of number of people per car before and after the introduction of the car-sharing scheme.

When calculating the expectation and variance of a discrete probability distribution, you will find it helpful to set your work out systematically in a table.

			(a)	(b)
r	p_r	$r p_r$	$r^2 p_r$	$(r - \mu)^2 p_r$
1	0.35	0.35	0.35	0.35
2	0.375	0.75	1.5	0
3	0.205	0.615	1.845	0.205
4	0.065	0.26	1.04	0.26
5	0.005	0.025	0.125	0.045
Totals	$\Sigma p_r = 1$	$\mu = E(X) = 2$	**4.86**	**Var(X) = 0.86**

In this case:

$$E(X) = \mu = \sum r p_r$$
$$= 1 \times 0.35 + 2 \times 0.375 + 3 \times 0.205 + 4 \times 0.065 + 5 \times 0.005$$
$$= 2$$

And *either* from **(a)**

This is μ.

$$\mathrm{Var}(X) = \sigma^2 = \sum r^2 p_r - \left[\sum r p_r\right]^2$$
$$= 1^2 \times 0.35 + 2^2 \times 0.375 + 3^2 \times 0.205 + 4^2 \times 0.065$$
$$+ 5^2 \times 0.005 - 2^2$$
$$= 4.86 - 4$$
$$= 0.86$$

or from **(b)**

$$\mathrm{Var}(X) = \sigma^2 = \sum (r - \mu)^2 p_r$$
$$= (1 - 2)^2 \times 0.35 + (2 - 2)^2 \times 0.375 + (3 - 2)^2 \times 0.205$$
$$+ (4 - 2)^2 \times 0.065 + (5 - 2)^2 \times 0.005$$
$$= 0.86$$

The equivalence of the two methods is proved in Appendix 2 at www.hoddereducation.com/cambridgeextras.

In practice, method **(a)** is to be preferred since the computation is usually easier, especially when the expectation is other than a whole number.

> ## ACTIVITY 4.2
>
> Carry out similar calculations for the expectation and variance of the probability distribution *before* the car-sharing experiment using the data on page 109. Using these two statistics, judge the success or otherwise of the scheme.

Example 4.3

The discrete random variable X has the following probability distribution:

r	0	1	2	3
p_r	0.2	0.3	0.4	0.1

Find:

(i) $E(X)$
(ii) $E(X^2)$
(iii) $Var(X)$

using

(a) $E(X^2) - \mu^2$
(b) $E([X - \mu]^2)$.

Solution

			(a)	(b)
r	p_r	rp_r	$r^2 p_r$	$(r - \mu)^2 p_r$
0	0.2	0	0	0.392
1	0.3	0.3	0.3	0.048
2	0.4	0.8	1.6	0.144
3	0.1	0.3	0.9	0.256
Totals	1	1.4	2.8	0.84

(i) $E(X) = \mu = \sum r\, p_r$
$$= 0 \times 0.2 + 1 \times 0.3 + 2 \times 0.4 + 3 \times 0.1 = 1.4$$

(ii) $E(X^2) = \sum r^2 p_r$
$$= 0 \times 0.2 + 1 \times 0.3 + 4 \times 0.4 + 9 \times 0.1 = 2.8$$

(iii) (a) $Var(X) = E(X^2) - \mu^2$
$$= 2.8 - 1.4^2 = 0.84$$

(b) $Var(X) = E([X - \mu]^2)$
$$= \sum (r - \mu)^2 p_r$$
$$= (0 - 1.4)^2 \times 0.2 + (1 - 1.4)^2 \times 0.3 + (2 - 1.4)^2 \times 0.4$$
$$+ (3 - 1.4)^2 \times 0.1$$
$$= 0.392 + 0.048 + 0.144 + 0.256 = 0.84$$

Notice that the two methods of calculating the variance in part (iii) give the same result, since one formula is just an algebraic rearrangement of the other.

> Look carefully at both methods for calculating the variance.
> › Are there any situations where one method might be preferred to the other?

As well as being able to carry out calculations for the expectation and variance you are often required to solve problems in context. The following example illustrates this idea.

Example 4.4

Laura buys one litre of mango juice on three days out of every four and none on the fourth day. A litre of mango juice costs 40c. Let X represent her weekly juice bill.

(i) Find the probability distribution of her weekly juice bill.
(ii) Find the mean (μ) and standard deviation (σ) of her weekly juice bill.
(iii) Find:
  (a) $P(X > \mu + \sigma)$
  (b) $P(X < \mu - \sigma)$.

Solution

(i) The pattern repeats every four weeks.

M	Tu	W	Th	F	Sa	Su	Number of litres	Juice bill
✓	✓	✓	✗	✓	✓	✓	6	$2.40
✗	✓	✓	✓	✗	✓	✓	5	$2.00
✓	✗	✓	✓	✓	✗	✓	5	$2.00
✓	✓	✗	✓	✓	✓	✗	5	$2.00

Tabulating the probability distribution for X gives the following.

r ($)	2.00	2.40
$P(X = r)$	0.75	0.25

(ii) $E(X) = \mu = \sum rP(X = r)$
$= 2 \times 0.75 + 2.4 \times 0.25$
$= 2.1$

$\text{Var}(X) = \sigma^2 = E(X^2) - \mu^2$
$= 4 \times 0.75 + 5.76 \times 0.25 - 2.1^2$
$= 0.03$

$\Rightarrow \quad \sigma = \sqrt{0.03} = 0.17$ (correct to 2 s.f.)

Hence her mean weekly juice bill is $2.10, with a standard deviation of about 17 cents.

(iii) (a) $P(X > \mu + \sigma) = P(X > 2.27) = 0.25$
(b) $P(X < \mu - \sigma) = P(X < 1.93) = 0$

1 Find by calculation the expectation of the outcome with the following probability distribution.

Outcome	1	2	3	4	5
Probability	0.1	0.2	0.4	0.2	0.1

2 The probability distribution of the discrete random variable X is given by

$$P(X = r) = \frac{2r - 1}{16} \quad \text{for } r = 1, 2, 3, 4$$

$$P(X = r) = 0 \quad \text{otherwise.}$$

(i) Find $E(X) = \mu$.

(ii) Find $P(X < \mu)$.

3 (i) A discrete random variable X can take only the values 4 and 5, and has expectation 4.2.

By letting $P(X = 4) = p$ and $P(X = 5) = 1 - p$, solve an equation in p and so find the probability distribution of X.

(ii) A discrete random variable Y can take only the values 50 and 100. Given that $E(Y) = 80$, write out the probability distribution of Y.

4 The random variable X is given by the sum of the scores when two ordinary dice are thrown.

(i) Use the shape of the distribution to find $E(X) = \mu$.
Confirm your answer by calculation.

(ii) Calculate $Var(X) = \sigma^2$.

(iii) Find the values of the following.

(a) $P(X < \mu)$ (b) $P(X > \mu + \sigma)$ (c) $P(|X - \mu| < 2\sigma)$

5 The random variable Y is given by the absolute difference between the scores when two ordinary dice are thrown.

(i) Find $E(Y)$ and $Var(Y)$.

(ii) Find the values of the following.

(a) $P(Y > \mu)$ (b) $P(Y > \mu + 2\sigma)$

CP **6** Three fair coins are tossed. Let X represent the number of tails.

(i) Find $E(X)$. Show that this is equivalent to $3 \times \frac{1}{2}$.

(ii) Find $Var(X)$. Show that this is equivalent to $3 \times \frac{1}{4}$.

If instead ten fair coins are tossed, let Y represent the number of tails.

(iii) Write down the values of $E(Y)$ and $Var(Y)$.

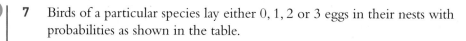

7 Birds of a particular species lay either 0, 1, 2 or 3 eggs in their nests with probabilities as shown in the table.

Number of eggs	0	1	2	3
Probability	0.25	0.35	0.30	k

Find:
(i)   the value of k
(ii) the expected number of eggs laid in a nest
(iii) the standard deviation of the number of eggs laid in a nest.

8 An electronic device produces an output of 0, 1 or 3 volts, with probabilities $\frac{1}{2}, \frac{1}{3}$ and $\frac{1}{6}$ respectively. The random variable X denotes the result of adding the outputs for two such devices, which act independently.

(i)   Show that $P(X = 4) = \frac{1}{9}$.

(ii) Tabulate all the possible values of X with their corresponding probabilities.

(iii) Hence calculate $E(X)$ and $Var(X)$, giving your answers as fractions in their lowest terms.

9 Bob earns \$80 per day, Monday to Friday inclusive. He works every alternate Saturday for which he earns 'time and a half' and every fourth Sunday, for which he is paid 'double time'.

(i)   By considering a typical four-week period of 28 days, find the probability distribution for his *daily* wage.

(ii) Calculate the expectation and variance of his *daily* wage.

(iii) Show that there are two possible patterns Bob could work over a typical four-week period, depending on which Saturdays and Sunday he works. Hence find the expectation and variance of his *weekly* wage under either pattern.

10 The probability distribution of the discrete random variable X is shown in the table below.

x	-3	-1	0	4
$P(X = x)$	a	b	0.15	0.4

Given that $E(X) = 0.75$, find the values of a and b.

Cambridge International AS & A Level Mathematics
9709 Paper 61 Q1 June 2010

11 Every day Eduardo tries to phone his friend. Every time he phones there is a 50% chance that his friend will answer. If his friend answers, Eduardo does not phone again on that day. If his friend does not answer, Eduardo tries again in a few minutes' time. If his friend has not answered after 4 attempts, Eduardo does not try again on that day.

(i) Draw a tree diagram to illustrate this situation.

(ii) Let X be the number of unanswered phone calls made by Eduardo on a day. Copy and complete the table showing the probability distribution of X.

x	0	1	2	3	4
$P(X = x)$		$\frac{1}{4}$			

(iii) Calculate the expected number of unanswered phone calls on a day.

Cambridge International AS & A Level Mathematics
9709 Paper 6 Q6 June 2008

12 Gohan throws a fair tetrahedral die with faces numbered 1, 2, 3, 4. If she throws an even number then her score is the number thrown. If she throws an odd number then she throws again and her score is the sum of both numbers thrown. Let the random variable X denote Gohan's score.

(i) Show that $P(X = 2) = \frac{5}{16}$.

(ii) The table below shows the probability distribution of X.

x	2	3	4	5	6	7
$P(X = x)$	$\frac{5}{16}$	$\frac{1}{16}$	$\frac{3}{8}$	$\frac{1}{8}$	$\frac{1}{16}$	$\frac{1}{16}$

Calculate $E(X)$ and $Var(X)$.

Cambridge International AS & A Level Mathematics
9709 Paper 6 Q2 June 2009

13 The probability distribution of the random variable X is shown in the following table.

x	-2	-1	0	1	2	3
$P(X = x)$	0.08	p	0.12	0.16	q	0.22

The mean of X is 1.05.

(i) Write down two equations involving p and q and hence find the values of p and q.

(ii) Find the variance of X.

Cambridge International AS & A Level Mathematics
9709 Paper 61 Q2 November 2009

14 The random variable X takes the values $-2, 0$ and 4 only. It is given that $P(X = -2) = 2p$, $P(X = 0) = p$ and $P(X = 4) = 3p$.

(i) Find p.

(ii) Find $E(X)$ and $Var(X)$.

Cambridge International AS & A Level Mathematics
9709 Paper 6 Q2 November 2007

15 A fair die has one face numbered 1, one face numbered 3, two faces numbered 5 and two faces numbered 6.

(i) Find the probability of obtaining at least 7 odd numbers in 8 throws of the die.

The die is thrown twice. Let X be the sum of the two scores. The following table shows the possible values of X.

			Second throw				
		1	3	5	5	6	6
First throw	1	2	4	6	6	7	7
	3	4	6	8	8	9	9
	5	6	8	10	10	11	11
	5	6	8	10	10	11	11
	6	7	9	11	11	12	12
	6	7	9	11	11	12	12

(ii) Draw up a table showing the probability distribution of X.

(iii) Calculate $E(X)$.

(iv) Find the probability that X is greater than $E(X)$.

Cambridge International AS & A Level Mathematics
9709 Paper 6 Q7 November 2008

KEY POINTS

1 For a discrete random variable, X, which can take only the values $r_1, r_2, \ldots, r_n$, with probabilities $p_1, p_2, \ldots, p_n$ respectively:

$$p_1 + p_2 + \ldots + p_n = \sum_{i=1}^{n} p_i = \sum_{i=1}^{n} P(X = r_i) = p_r = 1; \; p_i \geqslant 0$$

2 A discrete probability distribution is best illustrated by a vertical line chart.

• The expectation $= E(X) = \mu = \sum r P(x = r) = \sum r p_r$

• The variance, where σ is the standard deviation, is

$$Var(X) = \sigma^2 = E(X - \mu)^2 = \sum (r - \mu)^2 p_r$$

or $\;Var(X) = \sigma^2 = E(X^2) - [E(X)]^2 = \sum r^2 p_r - \left[\sum r p_r\right]^2$

3   Another common notation is to denote the values the variable may take by X.

  - The expectation = $E(X) = \sum xp$
  - The variance = $Var(X) = \sum x^2 p - [E(X)]^2$

LEARNING OUTCOMES

Now that you have finished this chapter, you should be able to

- understand the terms:
  - probability distribution
  - discrete random variable
  - expectation
- draw a probability distribution table
- calculate expectation and variance
- solve problems involving discrete random variables.

5

Permutations and combinations

The human brain, it has to be said, is the most complexly organised structure in the universe. There are 100 billion neurons in the adult human brain and each makes contacts with 1000 to 10 000 other neurons. Based on this information it has been calculated that the number of permutations and combinations of brain activity exceeds the number of elementary particles in the universe.
V. S. Ramachandran (1951–)

ProudMum

My son is a genius!
I gave Oscar five bricks and straightaway he did this! Is it too early to enrol him with MENSA?

> ❓ What is the probability that Oscar chose the bricks at random and just happened by chance to get them in the right order?

There are two ways of looking at the situation. You can think of Oscar selecting the five bricks as five events, one after another. Alternatively, you can think of 1, 2, 3, 4, 5 as one outcome out of several possible outcomes and work out the probability that way.

Five events

Look at the diagram.

▲ **Figure 5.1**

If Oscar had actually chosen them at random:

the probability of first selecting 1 is $\frac{1}{5}$

the probability of next selecting 2 is $\frac{1}{4}$

the probability of next selecting 3 is $\frac{1}{3}$

the probability of next selecting 4 is $\frac{1}{2}$

> 1 correct choice from 4 remaining bricks.

then only 5 remains so the probability of selecting it is 1.

So the probability of getting the correct numerical sequence at random is

$$\frac{1}{5} \times \frac{1}{4} \times \frac{1}{3} \times \frac{1}{2} \times 1 = \frac{1}{120}.$$

Outcomes

How many ways are there of putting five bricks in a line?

To start with there are five bricks to choose from, so there are five ways of choosing brick 1. Then there are four bricks left and so there are four ways of choosing brick 2. And so on.

The total number of ways is

5	×	4	×	3	×	2	×	1	= 120.
Brick 1		Brick 2		Brick 3		Brick 4		Brick 5	

Only one of these is the order 1, 2, 3, 4, 5, so the probability of Oscar selecting it at random is $\frac{1}{120}$.

> Number of possible outcomes.

> ❯ Do you agree with Oscar's mother that he is a child prodigy, or do you think it was just by chance that he put the bricks down in the right order?
>
> ❯ What further information would you want to be convinced that he is a budding genius?

5.1 Factorials

In the last example you saw that the number of ways of placing five different bricks in a line is $5 \times 4 \times 3 \times 2 \times 1$. This number is called 5 **factorial** and is written 5!. You will often meet expressions of this form.

In general the number of ways of placing n different objects in a line is $n!$, where $n! = n \times (n - 1) \times (n - 2) \times ... \times 3 \times 2 \times 1$.

n must be a positive integer.

Example 5.1

Calculate 7!

Solution

$7! = 7 \times 6 \times 5 \times 4 \times 3 \times 2 \times 1 = 5040$

Some typical relationships between factorial numbers are illustrated below:

$10! = 10 \times 9!$ or in general $n! = n \times [(n - 1)!]$

$10! = 10 \times 9 \times 8 \times 7!$ or in general $n! = n \times (n - 1) \times (n - 2) \times [(n - 3)!]$

These are useful when simplifying expressions involving factorials.

Example 5.2

Calculate $\dfrac{5!}{3!}$

Solution

$\dfrac{5!}{3!} = \dfrac{5 \times 4 \times 3!}{3!} = 5 \times 4 = 20$

Example 5.3

Calculate $\dfrac{7! \times 5!}{3! \times 4!}$

Solution

$\dfrac{7! \times 5!}{3! \times 4!} = \dfrac{7 \times 6 \times 5 \times 4 \times 3! \times 5 \times 4!}{3! \times 4!}$

$= 7 \times 6 \times 5 \times 4 \times 5 = 4200$

Example 5.4

Write $37 \times 36 \times 35$ in terms of factorials only.

Solution

$37 \times 36 \times 35 = \dfrac{37 \times 36 \times 35 \times 34!}{34!}$

$= \dfrac{37!}{34!}$

Example 5.5

(i) Find the number of ways in which all five letters in the word GREAT can be arranged.

(ii) In how many of these arrangements are the letters A and E next to each other?

Solution

(i) There are five choices for the first letter (G, R, E, A or T). Then there are four choices for the next letter, then three for the third letter and so on. So the number of arrangements of the letters is

$$5 \times 4 \times 3 \times 2 \times 1 = 5! = 120$$

(ii) The E and the A are to be together, so you can treat them as a single letter.

So there are four choices for the first letter (G, R, EA or T), three choices for the next letter and so on.

So the number of arrangements of these four 'letters' is

$$4 \times 3 \times 2 \times 1 = 4! = 24$$

However | EA | G | R | T |

is different from | AE | G | R | T |

So each of the 24 arrangements can be arranged into two different orders.

The total number of arrangements with the E and A next to each other is

$$2 \times 4! = 48$$

> ### Note
> The total number of ways of arranging the letters with the A and the E apart is
> $$120 - 48 = 72$$

Sometimes a question will ask you to deal with repeated letters.

Example 5.6

Find the number of ways in which all five letters in the word GREET can be arranged.

Solution

There are $5! = 120$ arrangements of five letters.

However, GREET has two repeated letters and so some of these arrangements are really the same.

For example, | E | E | G | R | T |

is the same as | E | E | G | R | T |

→

The two Es can be arranged in 2! = 2 ways, so the total number of arrangements is

$$\frac{5!}{2!} = 60$$

Example 5.7

How many different arrangements of the letters in the word MATHEMATICAL are there?

Solution

There are 12 letters, so there are 12! = 479 001 600 arrangements.

However, there are repeated letters and so some of these arrangements are the same.

For example,

| M | A | T | H | E | M | A | T | I | C | A | L |

| M | A | T | H | E | M | A | T | I | C | A | L |

and

| M | A | T | H | E | M | A | T | I | C | A | L |

are the same.

In fact, there are 3! = 6 ways of arranging the As.

So the total number of arrangements of

| M | A | T | H | E | M | A | T | I | C | A | L | is

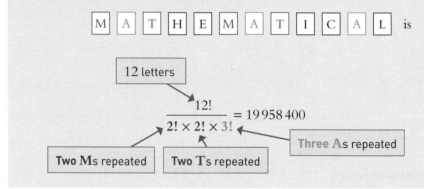

$$\frac{12!}{2! \times 2! \times 3!} = 19\,958\,400$$

Example 5.7 illustrates how to deal with repeated objects. You can generalise from this example to obtain the following:

▸▸ The number of distinct arrangements of n objects in a line, of which p are identical to each other, q others are identical to each other, r of a third type are identical, and so on is $\dfrac{n!}{p!q!r!\dots}$.

1 Calculate (i) 8! (ii) $\dfrac{8!}{6!}$ (iii) $\dfrac{5! \times 6!}{7! \times 4!}$

2 Simplify (i) $\dfrac{(n-1)!}{n!}$ (ii) $\dfrac{(n-1)!}{(n-2)!}$

3 Simplify (i) $\dfrac{(n+3)!}{(n+1)!}$ (ii) $\dfrac{n!}{(n-2)!}$

4 Write in factorial notation.

(i) $\dfrac{8 \times 7 \times 6}{5 \times 4 \times 3}$ (ii) $\dfrac{15 \times 16}{4 \times 3 \times 2}$ (iii) $\dfrac{(n+1)n(n-1)}{4 \times 3 \times 2}$

5 Factorise (i) $7! + 8!$ (ii) $n! + (n+1)!$

6 How many different four-letter words can be formed from the letters A, B, C and D if letters cannot be repeated? (The words do not need to mean anything.)

7 How many different ways can eight books be arranged in a row on a shelf?

8 In a motoring rally there are six drivers. How many different ways are there for the six drivers to finish?

9 In a 60-metre hurdles race there are five runners, one from each of the nations Austria, Belgium, Canada, Denmark and England.

(i) How many different finishing orders are there?

(ii) What is the probability of predicting the finishing order by choosing first, second, third, fourth and fifth at random?

10 Chenglei has an MP3 player which can play tracks in 'shuffle' mode. If an album is played in 'shuffle' mode the tracks are selected in a random order with a different track selected each time until all the tracks have been played.

Chenglei plays a 14-track album in 'shuffle' mode.

(i) In how many different orders could the tracks be played?

(ii) What is the probability that 'shuffle' mode will play the tracks in the normal set order listed on the album?

11 In a 'Goal of the season' competition, participants are asked to rank ten goals in order of quality.

The organisers select their 'correct' order at random. Anybody who matches their order will be invited to join the television commentary team for the next international match.

(i) What is the probability of a participant's order being the same as that of the organisers?

(ii) Five million people enter the competition. How many people would be expected to join the commentary team?

12 The letters O, P, S and T are placed in a line at random. What is the probability that they form a word in the English language?

13 Find how many arrangements there are of the letters in each of these words.

(i) EXAM (ii) MATHS (iii) CAMBRIDGE

(iv) PASS (v) SUCCESS (vi) STATISTICS

14 How many arrangements of the word ACHIEVE are there if:

(i) there are no restrictions on the order the letters are to be in

(ii) the first letter is an A

(iii) the letters A and I are to be together

(iv) the letters C and H are to be apart?

INVESTIGATIONS

1 Solve the inequality $n! > 10^m$ for each of the cases $m = 3, 4, 5$.

2 In how many ways can you write 42 using factorials only?

3 (i) There are 4! ways of placing the four letters S, T, A, R in a line, if each of them must appear exactly once. How many ways are there if each letter may appear any number of times (i.e. between 0 and 4)? Formulate a general rule.

(ii) There are 4! ways of placing the letters S, T, A, R in line. How many ways are there of placing in line the letters:

(a) S, T, A, A (b) S, T, T, T?

Formulate a general rule for dealing with repeated letters.

5.2 Permutations

I should be one of the judges! When I heard the 16 songs in the competition, I knew which ones I thought were the best three. Last night they announced the results and I had picked the same three songs in the same order as the judges!

Joyeeta

What is the probability of Joyeeta's result?

The winner can be chosen in 16 ways.
The second song can be chosen in 15 ways.
The third song can be chosen in 14 ways.

Thus the total number of ways of placing three songs in the first three positions is $16 \times 15 \times 14 = 3360$. So the probability that Joyeeta's selection is correct is $\frac{1}{3360}$.

In this example attention is given to the order in which the songs are placed. The solution required a **permutation** of three objects from sixteen.

In general the number of permutations, ^nP_r, of r objects from n is given by

$$^nP_r = n \times (n-1) \times (n-2) \times ... \times (n-r+1).$$

This can be written more compactly as

» $^nP_r = \dfrac{n!}{(n-r)!}$

| Example 5.8 | Six people go to the cinema. They sit in a row with ten seats. Find how many ways can this be done if: |

5

(i)   they can sit anywhere

(ii)   all the empty seats are next to each other.

Solution

(i)   The first person to sit down has a choice of ten seats.
       The second person to sit down has a choice of nine seats.
       The third person to sit down has a choice of eight seats.

       ...

       The sixth person to sit down has a choice of five seats.

       So the total number of arrangements is $10 \times 9 \times 8 \times 7 \times 6 \times 5 = 151\,200$. This is a permutation of six objects from ten, so a quicker way to work this out is

       $$\text{number of arrangements} = {}^{10}P_6 = 151\,200$$

(ii)   Since all four empty seats are to be together you can consider them to be a single 'empty seat', albeit a large one!
       So there are seven seats to seat six people.
       So the number of arrangements is ${}^{7}P_6 = 5040$

5.3 Combinations

It is often the case that you are not concerned with the order in which items are chosen, only with which ones are picked.

A maths teacher is playing a game with her students. Each student selects six numbers out of a possible 49 (numbers 1, 2, . . . , 49). The maths teacher then uses a random number machine to generate six numbers. If a student's numbers match the teacher's numbers then they win a prize.

> ❯ You have the six winning numbers. Does it matter in which order the machine picked them?

The teacher says that the probability of an individual student picking the winning numbers is about 1 in 14 million. How can you work out this figure?

The key question is, how many ways are there of choosing six numbers out of 49?

If the order mattered, the answer would be ${}^{49}P_6$, or $49 \times 48 \times 47 \times 46 \times 45 \times 44$.

However, the order does not matter. The selection 1, 3, 15, 19, 31 and 48 is the same as 15, 48, 31, 1, 19, 3 and as 3, 19, 48, 1, 15, 31, and lots more. For each set of six numbers there are 6! arrangements that all count as being the same.

So, the number of ways of selecting six numbers, given that the order does not matter, is

$$\frac{49 \times 48 \times 47 \times 46 \times 45 \times 44}{6!}.$$

This is $\dfrac{^{49}P_6}{6!}$

This is called the number of **combinations** of 6 objects from 49 and is denoted by $^{49}C_6$.

> ➤ Show that $^{49}C_6$ can be written as $\dfrac{49!}{6! \times 43!}$.

Returning to the maths teacher's game, it follows that the probability of a student winning is $\dfrac{1}{^{49}C_6}$.

> ➤ Check that this is about 1 in 14 million.

This example shows a general result, that the number of ways of selecting r objects from n, when the order does not matter, is given by

$$^nC_r = \frac{n!}{r!(n-r)!} = \frac{^nP_r}{r!}$$

> ➤ How can you prove this general result?

Another common notation for ^nC_r is $\binom{n}{r}$. Both notations are used in this book to help you become familiar with both of them.

> ❗ The notation $\binom{n}{r}$ looks exactly like a column vector and so there is the possibility of confusing the two. However, the context should usually make the meaning clear.

Example 5.9

A School Governors' committee of five people is to be chosen from eight applicants. How many different selections are possible?

Solution

Number of selections $= \binom{8}{5} = \dfrac{8!}{5! \times 3!} = \dfrac{8 \times 7 \times 6}{3 \times 2 \times 1} = 56$

| Example 5.10 |

In how many ways can a committee of four people be selected from four applicants?

Solution

Common sense tells us that there is only one way to make the committee, that is by appointing all applicants. So $^4C_4 = 1$. However, if we work from the formula

$$^4C_4 = \frac{4!}{4! \times 0!} = \frac{1}{0!}$$

For this to equal 1 requires the convention that 0! is taken to be 1.

> ❯ Use the convention 0! = 1 to show that $^nC_0 = {}^nC_n = 1$ for all values of n.

5.4 The binomial coefficients

In the last section you met numbers of the form ^nC_r or $\binom{n}{r}$. These are called the binomial coefficients; the reason for this is explained in Appendix 3 at www.hoddereducation.com/cambridgeextras and in the next chapter.

> ▶ **ACTIVITY 5.1**
>
> Use the formula $\binom{n}{r} = \dfrac{n!}{r!(n-r)!}$ and the results $\binom{n}{0} = \binom{n}{n} = 1$ to check that the entries in this table, for $n = 6$ and 7, are correct.
>
> | r | 0 | 1 | 2 | 3 | 4 | 5 | 6 | 7 |
> |-----|---|---|----|----|----|----|---|---|
> | $n = 6$ | 1 | 6 | 15 | 20 | 15 | 6 | 1 | – |
> | $n = 7$ | 1 | 7 | 21 | 35 | 35 | 21 | 7 | 1 |

It is very common to present values of ^nC_r in a table shaped like an isosceles triangle, known as Pascal's triangle.

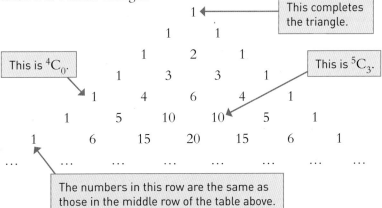

This completes the triangle.

This is $^4C_0$.

This is $^5C_3$.

The numbers in this row are the same as those in the middle row of the table above.

Pascal's triangle makes it easy to see two important properties of binomial coefficients.

1 Symmetry: $^{n}C_{r} = {}^{n}C_{n-r}$

If you are choosing 11 players from a pool of 15 possible players you can either name the 11 you have selected or name the 4 you have rejected. Similarly, every choice of r objects included in a selection from n distinct objects corresponds to a choice of $(n - r)$ objects which are excluded. Therefore $^{n}C_{r} = {}^{n}C_{n-r}$.

This provides a short cut in calculations when r is large. For example

$$^{100}C_{96} = {}^{100}C_{4} = \frac{100 \times 99 \times 98 \times 97}{1 \times 2 \times 3 \times 4} = 3\,921\,225.$$

It also shows that the list of values of $^{n}C_{r}$ for any particular value of n is unchanged by being reversed. For example, when $n = 6$ the list is the seven numbers 1, 6, 15, 20, 15, 6, 1.

2 Addition: $^{n+1}C_{r+1} = {}^{n}C_{r} + {}^{n}C_{r+1}$

Look at the entry 15 in the bottom row of Pascal's triangle, towards the right. The two entries above and either side of it are 10 and 5,

and 15 = 10 + 5. In this case $^{6}C_{4} = {}^{5}C_{3} + {}^{5}C_{4}$. This is an example of the general result that $^{n+1}C_{r+1} = {}^{n}C_{r} + {}^{n}C_{r+1}$. Check that all the entries in Pascal's triangle (except the 1s) are found in this way.

This can be used to build up a table of values of $^{n}C_{r}$ without much calculation. If you know all the values of $^{n}C_{r}$ for any particular value of n you can add pairs of values to obtain all the values of $^{n+1}C_{r}$, i.e. the next row, except the first and last, which always equal 1.

5.5 Using binomial coefficients to calculate probabilities

Example 5.11

A committee of 5 is to be chosen from a list of 14 people, 6 of whom are men and 8 women. Their names are to be put in a hat and then 5 drawn out.

What is the probability that this procedure produces a committee with no women?

Solution

The probability of an all-male committee of 5 people is given by

> There are 6 men.

$$\frac{\text{the number of ways of choosing 5 people out of 6}}{\text{the number of ways of choosing 5 people out of 14}} = \frac{^{6}C_{5}}{^{14}C_{5}} = \frac{6}{2002}$$

> There are 14 people.

$$= 0.003 \text{ to 3 d.p.}$$

GoByBus News

Help decide our new bus routes

The exact route for our new bus service is to be announced in April. Rest assured our service will run from Amli to Chatra via Bawal and will be extended to include Dhar once our new fleet of buses arrives in September. As local people know, there are several roads connecting these towns and we are keen to hear the views as to the most useful routes from our future passengers. Please post your views below!

RChowdhry

This consultation is a farce. The chance of getting a route that suits me is less than one in a hundred :(

Is RChowdhry right? How many routes are there from Amli to Dhar? Start by looking at the first two legs, Amli to Bawal and Bawal to Chatra.

There are three roads from Amli to Bawal and two roads from Bawal to Chatra. How many routes are there from Amli to Chatra passing through Bawal on the way?

Look at Figure 5.2.

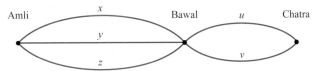

▲ **Figure 5.2**

The answer is $3 \times 2 = 6$ because there are three ways of doing the first leg, followed by two for the second leg. The six routes are

$$x - u \qquad y - u \qquad z - u$$
$$x - v \qquad y - v \qquad z - v.$$

There are also four roads from Chatra to Dhar. So each of the six routes from Amli to Chatral has four possible ways of going on to Dhar. There are now $6 \times 4 = 24$ routes. See Figure 5.3.

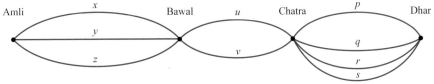

▲ **Figure 5.3**

They can be listed systematically as follows:

$$
\begin{array}{cccccc}
x-u-p & y-u-p & z-u-p & x-v-p & y-v-p & z-v-p \\
x-u-q & \cdots & \cdots & \cdots & \cdots & \cdots \\
x-u-r & \cdots & \cdots & \cdots & \cdots & \cdots \\
x-u-s & \cdots & \cdots & \cdots & \cdots & z-v-s
\end{array}
$$

In general, if there are a outcomes from experiment A, b outcomes from experiment B and c outcomes from experiment C then there are $a \times b \times c$ different possible combined outcomes from the three experiments.

> ❯ If GoByBus chooses its route at random, what is the probability that it will be the one R Chowdhry wants? Is the comment justified?
>
> ❯ In this example the probability was worked out by finding the number of possible routes. How else could it have been worked out?

Example 5.12

The manager of Avonford football squad wants to take a photo of the 13 players for the club magazine. She seats 6 players in the back row, then 5 best players in the middle row with the captain in the centre, and the 2 youngest players in the front row.

How many different ways can the players be organised for the photo?

Solution

The captain is sitting in the centre, so there are just 4 other players to seat in the middle row.

There are $6! = 720$ ways of arranging the 6 players in the back row.

There are $4! = 24$ ways of arranging the players in the middle row.

There are $2! = 2$ ways of seating the youngest players.

So altogether there are $6! \times 4! \times 2! = 720 \times 24 \times 2$

$$= 34\,560 \text{ different ways to seat the players}$$

Example 5.13

A cricket team consisting of 6 batsmen, 4 bowlers and 1 wicket-keeper is to be selected from a group of 18 cricketers comprising 9 batsmen, 7 bowlers and 2 wicket-keepers. How many different teams can be selected?

Solution

The batsmen can be selected in $^9C_6$ ways.

The bowlers can be selected in $^7C_4$ ways.

The wicket-keepers can be selected in $^2C_1$ ways.

Therefore total number of teams $= {}^9C_6 \times {}^7C_4 \times {}^2C_1$

$$= \frac{9!}{3! \times 6!} \times \frac{7!}{3! \times 4!} \times \frac{2!}{1! \times 1!}$$

$$= \frac{9 \times 8 \times 7}{3 \times 2 \times 1} \times \frac{7 \times 6 \times 5}{3 \times 2 \times 1} \times 2$$

$$= 5880$$

Example 5.14

In a dance competition, the panel of ten judges sit on the same side of a long table. There are three female judges.

(i)   How many different arrangements are there for seating the ten judges?

(ii) How many different arrangements are there if the three female judges all decide to sit together?

(iii) If the seating is at random, find the probability that the three female judges will **not** all sit together.

(iv) Four of the judges are selected at random to judge the final round of the competition. Find the probability that this final judging panel consists of two men and two women.

Solution

(i)   There are $10! = 3\,628\,800$ ways of arranging the judges in a line.

(ii) If the three female judges sit together then you can treat them as a single judge.

So there are eight judges and there are $8! = 40\,320$ ways of arranging the judges in a line.

However, there are $3! = 6$ ways of arranging the female judges.

So there are $3! \times 8! = 241\,920$ ways of arranging the judges so that all the female judges are together.

(iii) There are $3\,628\,800 - 241\,920 = 3\,386\,880$ ways of arranging the judges so that the female judges do not all sit together.

So the probability that the female judges do not all sit together is

$$\frac{3\,386\,880}{3\,628\,800} = 0.933 \quad \text{(to 3 s.f.)}.$$

(iv) The probability of selecting two men and two women on the panel of four is

$$\frac{{}^3C_2 \times {}^7C_2}{{}^{10}C_4} = \frac{3!}{1! \times 2!} \times \frac{7!}{5! \times 2!} \div \frac{10!}{6! \times 4!}$$

$$= 3 \times 21 \div 210$$

$$= 0.3$$

Exercise 5B

1 (i)   Find the values of   (a) $^6P_2$ (b) $^8P_4$ (c) $^{10}P_4$.

   (ii) Find the values of   (a) $^6C_2$ (b) $^8C_4$ (c) $^{10}C_4$.

   (iii) Show that, for the values of n and r in parts (i) and (ii),

   $$^nC_r = \frac{^nP_r}{r!}.$$

2 There are 15 runners in a camel race. What is the probability of correctly guessing the first three finishers in their finishing order?

3 A group of 5 computer programmers is to be chosen to form the night shift from a set of 14 programmers. In how many ways can the programmers be chosen if the 5 chosen must include the shift-leader who is one of the 14?

PS 4 My brother Mark decides to put together a rock band from amongst his year at school. He wants a lead singer, a guitarist, a keyboard player and a drummer. He invites applications and gets 7 singers, 5 guitarists, 4 keyboard players and 2 drummers. Assuming each person applies only once, in how many ways can Mark put the group together?

PS 5 A touring party of cricket players is made up of 5 players from each of India, Pakistan and Sri Lanka and 3 from Bangladesh.

   (i)   How many different selections of 11 players can be made for a team?

   (ii) In one match, it is decided to have 3 players from each of India, Pakistan and Sri Lanka and 2 from Bangladesh. How many different team selections can now be made?

PS 6 A committee of four is to be selected from ten candidates, six men and four women.

   (i)   In how many distinct ways can the committee be chosen?

   (ii) Assuming that each candidate is equally likely to be selected, determine the probabilities that the chosen committee contains:

   (a)   no women

   (b)   two men and two women.

PS 7 A committee of four is to be selected from five boys and four girls. The members are selected at random.

   (i)   How many different selections are possible?

   (ii) What is the probability that the committee will be made up of:

   (a)   all girls

   (b)   more boys than girls?

8 Baby Imran has a set of alphabet blocks. His mother often uses the blocks I, M, R, A and N to spell Imran's name.

(i) One day she leaves him playing with these five blocks. When she comes back into the room Imran has placed them in the correct order to spell his name. What is the probability of Imran placing the blocks in this order? (He is only 18 months old so he certainly cannot spell!)

(ii) A couple of days later she leaves Imran playing with all 26 of the alphabet blocks. When she comes back into the room she again sees that he has placed the five blocks I, M, R, A and N in the correct order to spell his name. What is the probability of him choosing the five correct blocks and placing them in this order?

9 (a) A football team consists of 3 players who play in a defence position, 3 players who play in a midfield position and 5 players who play in a forward position. Three players are chosen to collect a gold medal for the team. Find in how many ways this can be done

(i) if the captain, who is a midfield player, must be included, together with one defence and one forward player,

(ii) if exactly one forward player must be included, together with any two others.

(b) Find how many different arrangements there are of the nine letters in the words GOLD MEDAL

(i) if there are no restrictions on the order of the letters,

(ii) if the two letters D come first and the two letters L come last.

Cambridge International AS & A Level Mathematics
9709 Paper 6 Q7 June 2005

10 (a) Find how many different numbers can be made by arranging all nine digits of the number 223 677 888 if:

(i) there are no restrictions

(ii) the number made is an even number.

(b) Sandra wishes to buy some applications (apps) for her smartphone but she only has enough money for 5 apps in total. There are 3 train apps, 6 social network apps and 14 games apps available. Sandra wants to have at least 1 of each type of app. Find the number of different possible selections of 5 apps that Sandra can choose.

Cambridge International AS & A Level Mathematics
9709 Paper 61 Q7 June 2015

11

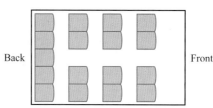

Back   Front

The diagram shows the seating plan for passengers in a minibus, which has 17 seats arranged in 4 rows. The back row has 5 seats and the other 3 rows have 2 seats on each side. 11 passengers get on the minibus.

(i)   How many possible seating arrangements are there for the 11 passengers?

(ii)   How many possible seating arrangements are there if 5 particular people sit in the back row?

Of the 11 passengers, 5 are unmarried and the other 6 consist of 3 married couples.

(iii) In how many ways can 5 of the 11 passengers on the bus be chosen if there must be 2 married couples and 1 other person, who may or may not be married?

Cambridge International AS & A Level Mathematics
9709 Paper 6 Q4 June 2006

12 Issam has 11 different CDs, of which 6 are pop music, 3 are jazz and 2 are classical.

(i)   How many different arrangements of all 11 CDs on a shelf are there if the jazz CDs are all next to each other?

(ii)   Issam makes a selection of 2 pop music CDs, 2 jazz CDs and 1 classical CD. How many different possible selections can be made?

Cambridge International AS & A Level Mathematics
9709 Paper 6 Q3 June 2008

13 A choir consists of 13 sopranos, 12 altos, 6 tenors and 7 basses. A group consisting of 10 sopranos, 9 altos, 4 tenors and 4 basses is to be chosen from the choir.

(i)   In how many different ways can the group be chosen?

(ii)   In how many ways can the 10 chosen sopranos be arranged in a line if the 6 tallest stand next to each other?

(iii) The 4 tenors and the 4 basses in the group stand in a single line with all the tenors next to each other and all the basses next to each other. How many possible arrangements are there if three of the tenors refuse to stand next to any of the basses?

Cambridge International AS & A Level Mathematics
9709 Paper 6 Q4 June 2009

14 A staff car park at a school has 13 parking spaces in a row. There are 9 cars to be parked.

(i) How many different arrangements are there for parking the 9 cars and leaving 4 empty spaces?

(ii) How many different arrangements are there if the 4 empty spaces are next to each other?

(iii) If the parking is random, find the probability that there will **not** be 4 empty spaces next to each other.

Cambridge International AS & A Level Mathematics
9709 Paper 6 Q3 November 2005

15 A builder is planning to build 12 houses along one side of a road. He will build 2 houses in style A, 2 houses in style B, 3 houses in style C, 4 houses in style D and 1 house in style E.

(i) Find the number of possible arrangements of these 12 houses.

(ii)

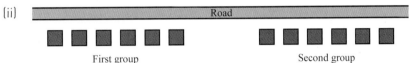

The 12 houses will be in two groups of 6 (see diagram). Find the number of possible arrangements if all the houses in styles A and D are in the first group and all the houses in styles B, C and E are in the second group.

(iii) Four of the 12 houses will be selected for a survey. Exactly one house must be in style B and exactly one house in style C. Find the number of ways in which these four houses can be selected.

Cambridge International AS & A Level Mathematics
9709 Paper 6 Q4 November 2008

16 (a) Find how many numbers between 5000 and 6000 can be formed from the digits 1, 2, 3, 4, 5 and 6

(i) if no digits are repeated

(ii) if repeated digits are allowed.

(b) Find the number of ways of choosing a school team of 5 pupils from 6 boys and 8 girls

(i) if there are more girls than boys in the team

(ii) if three of the boys are cousins and are either all in the team or all not in the team.

Cambridge International AS & A Level Mathematics
9709 Paper 61 Q5 November 2009

KEY POINTS

1 The number of ways of arranging n unlike objects in a line is $n!$

2 $n! = n \times (n-1) \times (n-2) \times (n-3) \times \ldots \times 3 \times 2 \times 1.$

3 The number of distinct arrangements of n objects in a line, of which p are identical to each other, q others are identical to each other, r of a third type are identical, and so on is

$$\frac{n!}{p!\,q!\,r!\ldots}.$$

4 The number of permutations of r objects from n is

$$^{n}\mathrm{P}_{r} = \frac{n!}{(n-r)!}.$$

5 The number of combinations of r objects from n is

$$^{n}\mathrm{C}_{r} = \frac{n!}{(n-r)!\,r!}.$$

This may also be written as $\binom{n}{r}$.

6 For permutations the order matters. For combinations it does not.

7 By convention $0! = 1$.

LEARNING OUTCOMES

Now that you have finished this chapter, you should be able to

■ understand the terms:
  - factorial
  - permutation
  - combination

■ solve problems involving arrangements where there are:
  - restrictions (e.g. X must not be next to Y)
  - repetitions (e.g. arrange letters in the word POOL)

■ solve problems involving selections where:
  - order matters (permutations)
  - order doesn't matter (combinations)
  - groupings (e.g. people sitting in rows)

■ understand the difference between permutations and combinations

■ use permutations and combinations to evaluate probabilities.

6 Discrete probability distributions

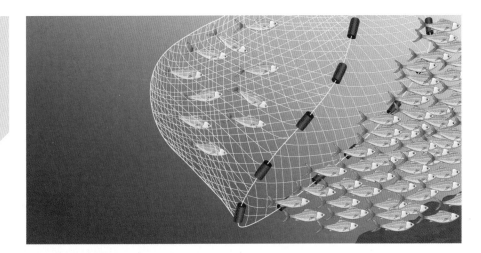

Innovation Stars – Blog

Samantha's great invention

Entrepreneur Samantha Weeks has done more than her bit to protect the environment. She has invented the first full spectrum LED bulb to operate on stored solar energy.

Now Samantha is out to prove that she is not only a clever scientist but a smart business woman as well. For Samantha is setting up her own factory to make and sell her bulbs.

Samantha admits there are still some technical problems …

Samantha Weeks hopes to make a big success of her light industry

Samantha's production process is not very good and there is a probability of 0.1 that any bulb will be substandard and so not last as long as it should.

She decides to sell her bulbs in packs of three. She believes that if one bulb in a pack is substandard the customers will not complain but that if two or more are substandard they will do so. She also believes that complaints should be kept down to no more than 2.5% of customers.

> ❯ Does Samantha meet her target?

Imagine a pack of Samantha's bulbs. There are eight different ways that good (G) and substandard (S) bulbs can be arranged in Samantha's packs, each with its associated probability.

Arrangement	Probability	Good	Substandard
G G G	$0.9 \times 0.9 \times 0.9 = 0.729$	3	0
G G S	$0.9 \times 0.9 \times 0.1 = 0.081$	2	1
G S G	$0.9 \times 0.1 \times 0.9 = 0.081$	2	1
S G G	$0.1 \times 0.9 \times 0.9 = 0.081$	2	1
G S S	$0.9 \times 0.1 \times 0.1 = 0.009$	1	2
S G S	$0.1 \times 0.9 \times 0.1 = 0.009$	1	2
S S G	$0.1 \times 0.1 \times 0.9 = 0.009$	1	2
S S S	$0.1 \times 0.1 \times 0.1 = 0.001$	0	3

Putting these results together gives this table.

Good	Substandard	Probability
3	0	0.729
2	1	0.243
1	2	0.027
0	3	0.001

So the probability of more than one substandard bulb in a pack is

$$0.027 + 0.001 = 0.028 \text{ or } 2.8\%.$$

This is slightly more than the 2.5% that Samantha regards as acceptable.

> ❯ What business advice would you give Samantha?

In this example we wrote down all the possible outcomes and found their probabilities one at a time. Even with just three bulbs this was repetitive. If Samantha had packed her bulbs in boxes of six it would have taken 64 lines to list them all. Clearly you need a more efficient approach.

You will have noticed that in the case of two good bulbs and one substandard, the probability is the same for each of the three arrangements in the box.

Arrangement	Probability	Good	Substandard
G G S	$0.9 \times 0.9 \times 0.1 = 0.081$	2	1
G S G	$0.9 \times 0.1 \times 0.9 = 0.081$	2	1
S G G	$0.1 \times 0.9 \times 0.9 = 0.081$	2	1

So the probability of this outcome is $3 \times 0.081 = 0.243$. The number 3 arises because there are three ways of arranging two good and one substandard bulb in the box. This is a result you have already met in the previous chapter but written slightly differently.

Example 6.1

How many different ways are there of arranging the letters GGS?

Solution

Since all the letters are either G or S, all you need to do is to count the number of ways of choosing the letter G two times out of three letters. This is

$$^3C_2 = \frac{3!}{2! \times 1!} = \frac{6}{2} = 3.$$

So what does this tell you? There was no need to list all the possibilities for Samantha's boxes of bulbs. The information could have been written down like this.

Good	Substandard	Expression	Probability
3	0	$^3C_3(0.9)^3$	0.729
2	1	$^3C_2(0.9)^2(0.1)^1$	0.243
1	2	$^3C_1(0.9)^1(0.1)^2$	0.027
0	3	$^3C_0(0.1)^3$	0.001

6.1 The binomial distribution

Samantha's light bulbs are an example of a common type of situation which is modelled by the binomial distribution. In describing such situations in this book, we emphasise the fact by using the word **trial** rather than the more general term **experiment**.

» You are conducting trials on random samples of a certain size, denoted by n.

» There are just two possible outcomes (in this case substandard and good). These are often referred to as *success* and *failure*.

» Both outcomes have fixed probabilities, the two adding to 1. The probability of success is usually called p, that of failure q, so $p + q = 1$.

» The probability of success is the same on each trial.

» The outcome of each trial is independent of any other trial.

You can then list the probabilities of the different possible outcomes as in the table above.

The method of the previous section can be applied more generally. You can call the probability of a substandard bulb p (instead of 0.1), the probability of a good bulb q (instead of 0.9) and the number of substandard bulbs in a packet of three, X.

Then the possible values of X and their probabilities are as shown in the table below.

r	0	1	2	3
$P(X = r)$	q^3	$3pq^2$	$3p^2q$	p^3

This package of values of X with their associated probabilities is called a **binomial probability distribution**, a special case of a discrete random variable.

If Samantha decided to put five bulbs in a packet the probability distribution would be as shown in the following table.

r	0	1	2	3	4	5
$P(X = r)$	q^5	$5pq^4$	$10p^2q^3$	$10p^3q^2$	$5p^4q$	p^5

10 is $^5C_2$.

The entry for $X = 2$, for example, arises because there are two 'successes' (substandard bulbs), giving probability p^2, and three 'failures' (good bulbs), giving probability q^3, and these can happen in $^5C_2 = 10$ ways. This can be written as $P(X = 2) = 10p^2q^3$.

If you are already familiar with the binomial theorem, you will notice that the probabilities in the table are the terms of the binomial expansion of $(q + p)^5$. This is why this is called a binomial distribution. Notice also that the sum of these probabilities is $(q + p)^5 = 1^5 = 1$, since $q + p = 1$, which is to be expected since the distribution covers all possible outcomes.

> **Note**
> The binomial theorem on the expansion of powers such as $(q + p)^n$ is covered in *Pure Mathematics 1*. The essential points are given in Appendix 3 at www.hoddereducation.com/cambridgeextras.

The general case

The general binomial distribution deals with the possible numbers of successes when there are n trials, each of which may be a success (with probability p) or a failure (with probability q); p and q are fixed positive numbers and $p + q = 1$. This distribution is denoted by $B(n, p)$. So, the original probability distribution for the number of substandard bulbs in Samantha's boxes of three is $B(3, 0.1)$.

For $B(n, p)$, the probability of r successes in n trials is found by the same argument as before. Each success has probability p and each failure has probability q, so the probability of r successes and $(n - r)$ failures in a particular order is p^rq^{n-r}. The positions in the sequence of n trials which the successes occupy can be chosen in ^nC_r ways. Therefore

$$P(X = r) = {}^nC_r p^r q^{n-r} \quad \text{for } 0 \leqslant r \leqslant n.$$

This can also be written as

$$p_r = \binom{n}{r} p^r (1 - p)^{n-r}.$$

The successive probabilities for $X = 0, 1, 2, ..., n$ are the terms of the binomial expansion of $(q + p)^n$.

Notes

1   The number of successes, X, is a variable which takes a restricted set of values ($X = 0, 1, 2, ..., n$) each of which has a known probability of occurring. This is an example of a *random variable*. Random variables are usually denoted by upper case letters, such as X, but the particular values they may take are written in lower case, such as r. To state that X has the binomial distribution $B(n, p)$ you can use the abbreviation $X \sim B(n, p)$, where the symbol $\sim$ means 'has the distribution'.

2   It is often the case that you use a theoretical distribution, such as the binomial, to describe a random variable that occurs in real life. This process is called modelling and it enables you to carry out relevant calculations. If the theoretical distribution matches the real-life variable perfectly, then the model is perfect. Usually, however, the match is quite good but not perfect. In this case the results of any calculations will not necessarily give a completely accurate description of the real-life situation. They may, nonetheless, be very useful.

Exercise 6A

1   A recovery ward in a maternity hospital has six beds. What is the probability that the mothers there have between them four girls and two boys? (You may assume that there are no twins and that a baby is equally likely to be a girl or a boy.)

2   A typist has a probability of 0.99 of typing a character correctly. He makes his mistakes at random. He types a sentence containing 200 characters. What is the probability that he makes exactly one mistake?

3   In a well-known game you have to decide which your opponent is going to choose: 'Paper', 'Stone' or 'Scissors'. If you guess entirely at random, what is the probability that you are right exactly 5 times out of 15?

4   There is a fault in a machine that makes microchips, with the result that only 80% of those it produces work. A random sample of eight microchips made by this machine is taken. What is the probability that exactly six of them work?

5 An airport is situated in a place where poor visibility (less than 800 m) can be expected 25% of the time. A pilot flies into the airport on ten different occasions.

(i) What is the probability that he encounters poor visibility exactly four times?

(ii) What other factors could influence the probability?

6 Three coins are tossed.

(i) What is the probability of all three showing heads?

(ii) What is the probability of two heads and one tail?

(iii) What is the probability of one head and two tails?

(iv) What is the probability of all three showing tails?

(v) Show that the probabilities for the four possible outcomes add up to 1.

(PS) **7** A coin is tossed ten times.

(i) What is the probability of it coming down heads five times and tails five times?

(ii) Which is more likely: exactly seven heads or more than seven heads?

8 In an election 30% of people support the Progressive Party. A random sample of eight voters is taken.

(i) What is the probability that it contains:
(a) 0   (b) 1   (c) 2   (d) at least 3
supporters of the Progressive Party?

(ii) Which is the most likely number of Progressive Party supporters to find in a sample size of eight?

(CP) **9** There are 15 children in a class.

(i) What is the probability that:
(a) 0   (b) 1   (c) 2   (d) at least 3
were born in January?

(ii) What assumption have you made in answering this question? How valid is this assumption in your view?

(CP) **10** Criticise this argument.

If you toss two coins they can come down three ways: two heads, one head and one tail, or two tails. There are three outcomes and so each of them must have probability one third.

6.2 The expectation and variance of B(n, p)

Example 6.2

The number of substandard bulbs in a packet of three of Samantha's bulbs is modelled by the random variable X where $X \sim B(3, 0.1)$.

(i) Find the expected frequencies of obtaining 0, 1, 2 and 3 substandard bulbs in 2000 packets.

(i) Find the mean number of substandard bulbs per packet.

Solution

(i) $P(X = 0) = 0.729$ (as on page 146), so the expected frequency of packets with no substandard bulbs is $2000 \times 0.729 = 1458$.

Similarly, the other expected frequencies are:

> Check:
> $1458 + 486 + 54$
> $+ 2 = 2000$

    for 1 substandard bulb: $2000 \times 0.243 = 486$
    for 2 substandard bulbs: $2000 \times 0.027 = 54$
    for 3 substandard bulbs: $2000 \times 0.001 = 2$.

(ii) The expected total of substandard bulbs in 2000 packets is

> This is also called the expectation.

$$0 \times 1458 + 1 \times 486 + 2 \times 54 + 3 \times 2 = 600.$$

Therefore the mean number of substandard bulbs per packet is $\frac{600}{2000} = 0.3$.

Notice in this example that to calculate the mean we have multiplied each probability by 2000 to get the frequency, multiplied each frequency by the number of faulty bulbs, added these numbers together and finally divided by 2000. Of course we could have obtained the mean with less calculation by just multiplying each number of faulty bulbs by its probability and then summing,

i.e. by finding $\sum_{r=0}^{3} rP(X = r)$. This is the standard method for finding an expectation, as you saw in Chapter 4.

Notice also that the mean or expectation of X is $0.3 = 3 \times 0.1 = np$. The result for the general binomial distribution is the same:

» if $X \sim B(n, p)$ then the expectation or mean of $X = \mu = np$.

This seems obvious: if the probability of success in each single trial is p, then the expected numbers of successes in n independent trials is np. However, since what seems obvious is not always true, a proper proof is required.

Let us take the case when $n = 5$. The distribution table for B$(5, p)$ is as on page 148, and the expectation of X is:

$$0 \times q^5 + 1 \times 5pq^4 + 2 \times 10p^2q^3 + 3 \times 10p^3q^2 + 4 \times 5p^4q + 5 \times p^5$$
$$= 5pq^4 + 20p^2q^3 + 30p^3q^2 + 20p^4q + 5p^5$$
$$= 5p(q^4 + 4pq^3 + 6p^2q^2 + 4p^3q + p^4)$$
$$= 5p(q + p)^4$$
$$= 5p$$

Take out the common factor $5p$.

Since $q + p = 1$.

The proof in the general case follows the same pattern: the common factor is now np, and the expectation simplifies to $np(q + p)^{n-1} = np$. The details are more fiddly because of the manipulations of the binomial coefficients.

Similarly, you can show that in this case the variance of X is given by $5pq$. This is an example of the general results that for a binomial distribution:

» mean $= \mu = np$

» variance, $\text{Var}(X) = \sigma^2 = npq = np(1 - p)$

» standard deviation $= \sigma = \sqrt{npq} = \sqrt{np(1 - p)}$.

> ## ACTIVITY 6.1
>
> If you want a challenge, write out the details of the proof that if $X \sim \text{B}(n, p)$ then the expectation of X is np.

6.3 Using the binomial distribution

Example 6.3

Which is more likely: that you get at least one 6 when you throw a die six times, or that you get at least two 6s when you throw it twelve times?

Solution

On a single throw of a die the probability of getting a 6 is $\frac{1}{6}$ and that of not getting a 6 is $\frac{5}{6}$.

So the probability distributions for the two situations required are B$(6, \frac{1}{6})$ and B$(12, \frac{1}{6})$ giving probabilities of:

$$1 - {}^6C_0\left(\frac{5}{6}\right)^6 = 1 - 0.335 = 0.665 \text{ (at least one 6 in six throws)}$$

and
$$1 - \left[{}^{12}C_0\left(\frac{5}{6}\right)^{12} + {}^{12}C_1\left(\frac{5}{6}\right)^{11}\left(\frac{1}{6}\right)\right] = 1 - (0.112 + 0.269)$$
$$= 0.619 \text{ (at least two 6s in 12 throws)}$$

So at least one 6 in six throws is somewhat more likely.

Example 6.4

Extensive research has shown that 1 person out of every 4 is allergic to a particular grass seed. A group of 20 university students volunteer to try out a new treatment.

(i) What is the expectation of the number of allergic people in the group?

(ii) What is the probability that:

    (a) exactly two

    (b) no more than two

    of the group are allergic?

(iii) How large a sample would be needed for the probability of it containing at least one allergic person to be greater than 99.9%?

(iv) What assumptions have you made in your answer?

Solution

This situation is modelled by the binomial distribution with $n = 20$, $p = 0.25$ and $q = 0.75$. The number of allergic people is denoted by X.

(i) Expectation $= np = 20 \times 0.25 = 5$ people.

(ii) $X \sim B(20, 0.25)$

    (a) $P(X = 2) = {}^{20}C_2(0.75)^{18}(0.25)^2 = 0.067$

    (b) $P(X \leqslant 2) = P(X = 0) + P(X = 1) + P(X = 2)$

$$= (0.75)^{20} + {}^{20}C_1(0.75)^{19}(0.25) + {}^{20}C_2(0.75)^{18}(0.25)^2$$
$$= 0.003 + 0.021 + 0.067$$
$$= 0.091$$

(iii) Let the sample size be n (people), so that $X \sim B(n, 0.25)$.

The probability that none of them is allergic is

$$P(X = 0) = (0.75)^n$$

and so the probability that at least one is allergic is

$$P(X \geqslant 1) = 1 - P(X = 0)$$
$$= 1 - (0.75)^n$$

So we need
$$1 - (0.75)^n > 0.999$$
$$(0.75)^n < 0.001$$
$$n \log 0.75 < \log 0.001$$
$$n > \log 0.001 \div \log 0.75$$
$$n > 24.01$$

> You meet logarithms in *Pure Mathematics 2 & 3*.

> log 0.75 is negative so you need to reverse the inequality.

So 25 people are required.

Notes

1   Although 24.01 is very close to 24 it would be incorrect to round down.
$1 - (0.75)^{24} = 0.998\,996\,6$ which is just less than 99.9%.

2   You can also use trial and improvement on a calculator to solve for n.

(iv) The assumptions made are:

» That the sample is random. This is almost certainly untrue. University students are nearly all in the 18–25 age range and so a sample of them cannot be a random sample of the whole population. They may well also be unrepresentative of the whole population in other ways. Volunteers are seldom truly random.

» That the outcome for one person is independent of that for another. This is probably true unless they are a group of friends from, say, an athletics team, where those with allergies are less likely to be members.

EXPERIMENT

Does the binomial distribution really work?

In the first case in Example 6.3, you threw a die six times (or six dice once each, which amounts to the same thing).

$X \sim \text{B}(6, \frac{1}{6})$ and this gives the probabilities in the following table.

Number of 6s	Probability
0	0.335
1	0.402
2	0.201
3	0.054
4	0.008
5	0.001
6	0.000

So if you carry out the experiment of throwing six dice 1000 times and record the number of 6s each time, you should get no 6s about 335 times, one 6 about 402 times, and so on. What does 'about' mean? How close an agreement can you expect between experimental and theoretical results?

You could carry out the experiment with dice, but it would be very tedious even if several people shared the work. Alternatively you could simulate the experiment on a spreadsheet using a random number generator.

Exercise 6B

1 In a game five dice are rolled together.

(i) What is the probability that:

(a) all five show 1

(b) exactly three show 1

(c) none of them shows 1?

(ii) What is the most likely number of times for 6 to show?

2 A certain type of sweet comes in eight colours: red, orange, yellow, green, blue, purple, pink and brown and these normally occur in equal proportions. Veronica's mother gives each of her children 16 of the sweets. Veronica says that the blue ones are much nicer than the rest and is very upset when she receives less than her fair share of them.

(i) How many blue sweets did Veronica expect to get?

(ii) What was the probability that she would receive fewer blue ones than she expected?

(iii) What was the probability that she would receive more blue ones than she expected?

3 Find the:

(i) mean

(ii) variance

of the following binomial distributions.

(a) $X \sim B(10, 0.25)$

(b) $X \sim B(10, 0.5)$

(c) $X \sim B(10, 0.75)$

4 In a particular area 30% of men and 20% of women are overweight and there are four men and three women working in an office there. Find the probability that there are:

(i) 0

(ii) 1

(iii) 2

overweight men

(iv) 0

(v) 1

(vi) 2

overweight women

(vii) exactly 2 overweight people in the office.

What assumption have you made in answering this question?

5 On her drive to work Stella has to go through four sets of traffic lights. She estimates that for each set the probability of her finding them red is $\frac{2}{3}$ and green $\frac{1}{3}$. (She ignores the possibility of them being amber.) Stella also estimates that when a set of lights is red she is delayed by one minute.

(i) Find the probability of:

(a) 0

(b) 1

(c) 2

(d) 3

sets of lights being against her.

(ii) Find the expected extra journey time due to waiting at lights.

6 Pepper moths are found in two varieties, light and dark. The proportion of dark moths increases with certain types of atmospheric pollution. At the time of the question 30% of the moths in a particular town are dark. A research student sets a moth trap and catches nine moths, four light and five dark.

(i) What is the probability of that result for a sample of nine moths?

(ii) Find the mean and variance of dark moths in samples of nine moths.

The next night the student's trap catches ten pepper moths.

(iii) What is the expected number of dark moths in this sample?

(iv) Find the probability that the actual number of dark moths in the sample is the same as the expected number.

7 (i) State three conditions which must be satisfied for a situation to be modelled by a binomial distribution.

George wants to invest some of his monthly salary. He invests a certain amount of this every month for 18 months. For each month there is a probability of 0.25 that he will buy shares in a large company, a probability of 0.15 that he will buy shares in a small company and a probability of 0.6 that he will invest in a savings account.

(ii) Find the probability that George will buy shares in a small company in at least 3 of these 18 months.

Cambridge International AS & A Level Mathematics
9709 Paper 61 Q3 June 2014

8 Biscuits are sold in packets of 18. There is a constant probability that any biscuit is broken, independently of other biscuits. The mean number of broken biscuits in a packet has been found to be 2.7. Find the probability that a packet contains between 2 and 4 (inclusive) broken biscuits.

Cambridge International AS & A Level Mathematics
9709 Paper 61 Q1 June 2011

9 In a certain mountainous region in winter, the probability of more than 20 cm of snow falling on any particular day is 0.21.

(i) Find the probability that, in any 7-day period in winter, fewer than 5 days have more than 20 cm of snow falling.

(ii) For 4 randomly chosen 7-day periods in winter, find the probability that exactly 3 of these periods will have at least 1 day with more than 20 cm of snow falling.

Cambridge International AS & A Level Mathematics
9709 Paper 61 Q4 June 2012

10 A box contains 300 discs of different colours. There are 100 pink discs, 100 blue discs and 100 orange discs. The discs of each colour are numbered from 0 to 99. Five discs are selected at random, one at a time, with replacement. Find:

(i) the probability that no orange discs are selected

(ii) the probability that exactly 2 discs with numbers ending in a 6 are selected

(iii) the probability that exactly 2 orange discs with numbers ending in a 6 are selected

(iv) the mean and variance of the number of pink discs selected.

Cambridge International AS & A Level Mathematics
9709 Paper 6 Q5 November 2005

11 The mean number of defective batteries in packs of 20 is 1.6. Use a binomial distribution to calculate the probability that a randomly chosen pack of 20 will have more than 2 defective batteries.

Cambridge International AS & A Level Mathematics
9709 Paper 61 Q1 November 2009

6.4 The geometric distribution

Asha is playing a game where you have to throw ten or more on two dice in order to start. She takes six throws to get a score of ten or more, and says that she has always been unlucky. One of the other players was successful on her first attempt.

> There are 36 possible outcomes when two dice are thrown. Those that give a total of 10 or more are 4,6; 5,5; 5,6; 6,4; 6,5; 6,6. So the probability of a total of 10 or more is $\frac{6}{36} = \frac{1}{6}$.

» The probability that Asha is successful on any attempt is $\frac{1}{6}$.

» The probability that Asha is unsuccessful on any particular attempt is therefore $\frac{5}{6}$.

» Asha has to have five failures followed by one success. So the probability that Asha is successful on her sixth attempt is $\left(\frac{5}{6}\right)^5 \times \frac{1}{6} = 0.0670$.

> Asha is successful on her sixth attempt so her results are F, F, F, F, F, S with probabilities $\frac{5}{6}, \frac{5}{6}, \frac{5}{6}, \frac{5}{6}, \frac{5}{6}, \frac{1}{6}$.

» So, in fact, it is about half as likely for Asha to succeed on her sixth attempt as it is on her first attempt.

The number of attempts that Asha takes to succeed is an example of a **geometric random variable**. The probability distribution is called the **geometric distribution**.

Example 6.5

Gina likes having an attempt at the coconut shy whenever she goes to the fair. From experience, she knows that the probability of her knocking over a coconut at any throw is $\frac{1}{3}$.

(i) Find the probability that Gina knocks over a coconut for the first time on her fifth attempt.

(ii) Find the probability that it takes Gina at most three attempts to knock over a coconut.

(iii) Given that Gina has already had four unsuccessful attempts, find the probability that it takes her another three attempts to succeed.

Solution

(i) Gina has to have 4 failures followed by 1 success. The probability of this

is $\left(\dfrac{2}{3}\right)^4 \times \dfrac{1}{3} = \dfrac{16}{243} = 0.0658$. ◄

> If Gina is successful on her fifth attempt her results are F, F, F, F, S with probabilities $\dfrac{2}{3}, \dfrac{2}{3}, \dfrac{2}{3}, \dfrac{2}{3}, \dfrac{1}{3}$.

Note
There is not much difference in the difficulty of these two methods if you want the probability of at most three attempts, but if you, for example, wanted the probability of at most 20 attempts, Method 2 would be far better as the calculation would take no more effort than in this case.

(ii) **Method 1**

This is the probability that she is successful on her first or second or third attempt. You can do similar calculations to the above:

$$\dfrac{1}{3} + \left(\dfrac{2}{3}\right)^1 \times \dfrac{1}{3} + \left(\dfrac{2}{3}\right)^2 \times \dfrac{1}{3} = \dfrac{1}{3} + \dfrac{2}{9} + \dfrac{4}{27} = \dfrac{19}{27} = 0.7037$$

Method 2

You can, instead, first work out the probability that Gina fails on all of her first three attempts and then subtract this from 1.

P (at most three attempts) = 1 − P (fails on all of first three attempts)

$$= 1 - \left(\dfrac{2}{3}\right)^3 = 1 - \dfrac{8}{27} = \dfrac{19}{27} = 0.7037$$

(iii) Unfortunately for Gina, however many unsuccessful attempts she has had makes no difference to how many more attempts she will need.

Thus the required probability is $\left(\dfrac{2}{3}\right)^2 \times \dfrac{1}{3} = \dfrac{4}{27} = 0.1481$.

This is another example of a geometric random variable. Part (iii) illustrates how the geometric distribution has 'no memory'.

For a geometric distribution to be appropriate, the following conditions must apply:

>> You are finding the number of trials it takes for the first success to occur.

>> On each trial, the outcomes can be classified as either success or failure.

> The probability of success is usually denoted by p and that of failure by q, so $p + q = 1$.

In addition, the following assumptions are needed if the geometric distribution is to be a good model and give reliable answers:

>> The outcome of each trial is independent of the outcome of any other trial.

>> The probability of success is the same on each trial.

In general, the geometric probability distribution $X \sim \text{Geo}(p)$ over the values $\{1, 2, 3, \dots\}$ is defined as follows:

> As with the binomial distribution, $q = 1 - p$.

$$P(X = r) = (1 - p)^{r-1}p \text{ or } q^{r-1}p \quad \text{for } r = 1, 2, 3\dots$$
$$P(X = r) = 0 \quad \text{otherwise.}$$

In Method 2 of part (ii) of Example 6.5, you saw that the probability of Gina failing on all of her first three attempts is $\left(\dfrac{2}{3}\right)^3$. More generally, a useful feature

Note

$P(X > r)$ is the probability that you take more than r attempts to succeed. The first r tries must therefore be failures. So $P(X > r) = (1-p)^r$.

of the geometric distribution is that $P(X > r) = (1-p)^r$, since this represents the probability that all of the first r attempts are failures.

The vertical line chart in Figure 6.1 illustrates the geometric distribution for $p = 0.3$.

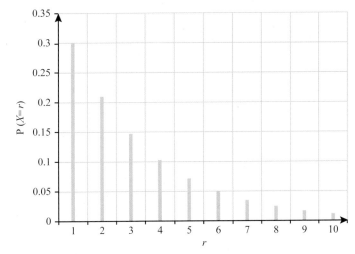

▲ **Figure 6.1**

For the geometric distribution with probability of success p

» $E(X) = \dfrac{1}{p}$

Example 6.6

In a communication network, messages are received one at a time and checked for errors. It is known that 6% of messages have an error in them. Errors in one message are independent of errors in any other message.

(i)   Find the probability that the first message that contains an error is the fifth message to be checked.

(ii)   Find the mean number of messages before the first that contains an error is found.

(iii)   Find the probability that there are no errors in the first ten messages.

(iv)   Find the most likely number of messages to be checked to find the first that contains an error.

Solution

In this situation, the distribution is geometric with probability 0.06, so Geo(0.06) so let $X \sim \text{Geo}(0.06)$.

(i) $P(X = 5) = 0.94^4 \times 0.06$
$= 0.0468$

(ii)   Mean $= \dfrac{1}{0.06}$
$= 16.7$ messages

(iii) $P(X > 10) = 0.94^{10}$
$= 0.539$

(iv) The probability of the first error being in the first message is 0.06.

The probability that it is in the second message is 0.94×0.06 and this is less than 0.06, and so on.

The probabilities get smaller each time.

So the most likely number is 1.

Example 6.7

The random variable $X \sim \text{Geo}(p)$. You are given that $E(X) = 5$.

(i) Find the value of p.

(ii) Find $P(X = 7)$.

(iii) Find $P(X = 10 \mid X > 7)$.

Solution

(i) $E(X) = \dfrac{1}{p} = 5$

$\Rightarrow p = 0.2$

(ii) $P(X = 7) = 0.8^6 \times 0.2$

$= 0.0524$

(iii) $P(X = 10 \mid X > 7) = P(X = 3)$

$= 0.8^2 \times 0.2$

$= 0.128$

Exercise 6C

1 A fair six-sided die is rolled. Find the probability that the first time a 6 comes up is

(i) on the first roll

(ii) on the third roll

(iii) before the third roll

(iv) after the third roll.

2 A fair three-sided spinner has sectors labelled 1, 2, 3. The spinner is spun until a 3 is scored. The number of spins required to get a 3 is denoted by X.

(i) Find $P(X = 1)$.

(ii) Find $P(X = 5)$.

(iii) Find $P(X > 5)$.

CP 3 A five-sided spinner has sectors labelled 2, 3, 4, 5 and 6. It is spun repeatedly and every time all the numbers are equally likely to come up when it is spun.

(i) Find the probability that

(a) on the first four spins it does not come up 2

(b) it comes up 2 before the fifth spin

(c)   on the first five spins it does not come up 2

(d)   the first time it comes up 2 is on the fifth spin.

(ii)   Find which three of your answers to part (i) add to 1 and explain why.

4   Two fair six-sided dice are rolled until a double comes up; that is, until they both show the same number.

(i)   What is the probability that on any roll of the dice a double comes up?

(ii)   Find the expected number of rolls that are needed to obtain the first double.

(iii)   What is the probability that the first double comes up on the sixth roll of the dice?

M   5   A game reserve runs safaris where people travel in a Jeep and look out for wild animals. Experience has shown that on one in four safaris a tiger is seen. The probability of seeing a tiger on any safari is independent of the probability of seeing one on any other safari.

Kamil goes on five safaris. Find the probability that

(i)   he does not see any tigers

(ii)   he sees at least one tiger

(iii)   the first tiger he sees is on his last safari

(iv)   he does not see any tigers on his first three safaris but does on his fourth and fifth safaris.

M   6   An archer is aiming for the bullseye on a target. The probability of the archer hitting the target on any attempt is 0.2, independent of any other attempt.

(i)   Find the probability that the archer hits the bullseye on her fifth attempt.

(ii)   Find the probability that the archer hits the bullseye for the first time on her fifth attempt.

(iii)   Find the probability that the archer takes at least 5 attempts to hit the bullseye.

(iv)   Find the mean of the number of attempts it takes for the archer to hit the bullseye.

(v)   The archer takes n shots at the bullseye. Find the smallest value of n for which there is a chance of 50% or more that she will hit the target at least once.

7   Fair four-sided dice with faces labelled 1, 2, 3, 4 are rolled, one at a time, until a 4 is scored.

(i)   Find the probability that the first 4 is rolled on the fifth attempt.

(ii)   Find the probability that it takes at least six attempts to roll the first 4.

(iii)   Given that a 4 is not rolled in the first six attempts, find the probability that a 4 is rolled within the next two attempts.

(iv)   Write down the mean number of attempts that it takes to roll a 4.

8 In a game show, a player is asked questions one after another until they get one wrong, after which it is the next player's turn. The probability that they get a question correct is 0.7, independent of any other question.

(i) Find the probability that the first player is asked a total of six questions.

(ii) Find the probability that the first player is asked at least six questions.

(iii) Find the average number of questions that the first player is asked.

(iv) Find the probability that the first two players are asked a total of five questions.

(M) **9** Hamish goes on a five-day salmon fishing holiday. The probability of catching a salmon on any day is $\frac{1}{3}$ and that is independent of what happens on any other day.

(i) What is the probability that he does not catch a salmon?

(ii) What is the probability that he catches his first salmon on the last day of his holiday?

(iii) What is the expectation of the number of days on which he catches a salmon?

(iv) What is the expectation of the number of days on which he does not catch a salmon?

(CP) **10** In order to start a board game, each player rolls fair six-sided dice, one at a time, until a 6 is obtained. Let X be the number of goes a player takes to start the game.

(i) Write down the distribution of X.

(ii) Find:

(a) $P(X = 4)$ (b) $P(X < 4)$ (c) $P(X > 4)$.

(iii) Given that $X = 3$, find the probability that the total score on all three of the dice is less than 10.

(iv) The game has two players and they each take turns at rolling the dice. Find the probability that:

(a) neither player has started within four rolls (two rolls each)

(b) both players have started within four rolls (two rolls each).

KEY POINTS

1 The binomial distribution may be used to model situations in which these conditions hold:

- you are conducting trials on random samples of a certain size, n
- in each trial there are two possible outcomes, often referred to as success and failure
- both outcomes have fixed probabilities, p and q, and $p + q = 1$.

2 For the binomial distribution to be a good model, these assumptions are needed:

- the probability of success is the same on each trial
- the outcome of each trial is independent of any other trial.

3   The probability that the number of successes, X, has the value r, is given by

$$P(X = r) = \binom{n}{r} p^r q^{n-r} = \binom{n}{r} p^r (1-p)^{n-r}$$

An alternative notation for $\binom{n}{r}$ is ^nC_r.

4   For $B(n, p)$:

- the expectation or mean of the number of successes, $E(X) = \mu = np$
- the variance, $Var(X) = \sigma^2 = npq = np(1-p)$
- the standard deviation, $\sigma = \sqrt{npq} = \sqrt{np(1-p)}$.

5   The geometric distribution may be used in situations in which

- there are two possible outcomes, often referred to as success and failure
- both outcomes have fixed probabilities, p and q, and $p + q = 1$
- you are finding the number of trials which it takes for the first success to occur.

6   For the geometric distribution to be a good model:

- the probability of success is constant
- the probability of success in any trial is independent of the outcome of any other trial.

7   For a geometric random variable X where $X \sim Geo(p)$:

- $P(X = r) = (1-p)^{r-1}p$ for $r = 1, 2, 3 \ldots$

8   For $Geo(p)$:

- $E(X) = \dfrac{1}{p}$

LEARNING OUTCOMES

Now that you have finished this chapter, you should be able to

- recognise situations that give rise to a binomial distribution, and express a binomial model in the form $B(n, p)$
- calculate probabilities using the binomial distribution
- find the mean and variance of a binomial distribution
- recognise situations under which the geometric distribution is likely to be an appropriate model
- calculate probabilities using a geometric distribution, including cumulative probabilities
- know and be able to use the mean of a geometric distribution.

7 The normal distribution

The
normal
law of error
stands out in the
experience of mankind
as one of the broadest
generalisations of natural
philosophy. It serves as the
guiding instrument in researches
in the physical and social sciences
and in medicine, agriculture and engineering.
It is an indispensable tool for the analysis and the
interpretation of the basic data obtained by observation and experiment.

W. J. Youden
(1900 – 1971)

> **To be and not be to, that is the answer.**
> *Piet Hein*
> *(1905–1996)*

UK Beanpole _____

Just had my height measured at the doctor's – I'm 194.3 cm. Can't be many around as tall as me!

> ❯ UK Beanpole is clearly exceptionally tall, but how much so?
>
> ❯ Is he one in a hundred, or a thousand or even a million? To answer that question you need to know the distribution of heights of adult British men.

The first point that needs to be made is that height is a continuous variable and not a discrete one. If you measure accurately enough it can take any value.

This means that it does not really make sense to ask 'What is the probability that somebody chosen at random has height exactly 194.3 cm?'. The answer is zero.

However, you can ask questions like 'What is the probability that somebody chosen at random has height between 194.25 cm and 194.35 cm?' and 'What is the probability that somebody chosen at random has height at least 194.3 cm?'. When the variable is continuous, you are concerned with a range of values rather than a single value.

Like many other naturally occurring variables, the heights of adult men may be modelled by a normal distribution, shown in Figure 7.1. You will see that this has a distinctive bell-shaped curve and is symmetrical about its middle. The curve is continuous as height is a continuous variable.

On Figure 7.1 area represents probability, so the shaded area to the right of 194.3 cm represents the probability that a randomly selected adult male is over 194.3 cm tall.

> ⚠ You should be aware that values given in tables are rounded. Consequently the final digit of an answer you obtain using tables may not be quite the same as if you have used the statistical function on your calculator instead. This table gives 4 figures so it is good practice to round your final answer to 3 significant figures.

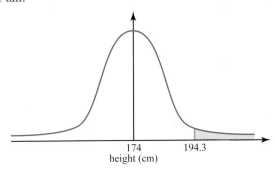

▲ **Figure 7.1**

Before you can start to find this area, you must know the mean and standard deviation of the distribution, in this case about 174 cm and 7 cm respectively.

So UK Beanpole's height is 194.3 cm − 174 cm = 20.3 cm above the mean, and that is

$$\frac{20.3}{7} = 2.9 \text{ standard deviations.}$$

The number of standard deviations beyond the mean, in this case 2.9, is denoted by the letter z. Thus the shaded area gives the probability of obtaining a value of $z \geqslant 2.9$.

You find this area by looking up the value of $\Phi(z)$ when $z = 2.9$ in a normal distribution table of $\Phi(z)$ as shown in Figure 7.2, and then calculating $1 - \Phi(z)$. (Φ is the Greek letter *phi*.)

z	0	1	2	3	4	5	6	7	8	9	1	2	3	4	5	6	7	8	9
															ADD				
0.0	0.5000	0.5040	0.5080	0.5120	0.5160	0.5199	0.5239	0.5279	0.5319	0.5359	4	8	12	16	20	24	28	32	36
0.1	0.5398	0.5438	0.5478	0.5517	0.5557	0.5596	0.5636	0.5675	0.5714	0.5753	4	8	12	16	20	24	28	32	36
0.2	0.5793	0.5832	0.5871	0.5910	0.5948	0.5987	0.6026	0.6064	0.6103	0.6141	4	8	12	15	19	23	27	31	35
0.3	0.6179	0.6217	0.6255	0.6293	0.6331	0.6368	0.6406	0.6443	06.480	0.6517	4	7	11	15	19	22	26	30	34
0.4	0.6554	0.6591	0.6628	0.6664	0.6700	0.6736	0.6772	0.6808	0.6844	0.6879	4	7	11	14	18	22	25	29	32
0.5	0.6915	0.6950	0.6985	0.7019	0.7054	0.7088	0.7123	0.7157	0.7190	0.7224	3	7	10	14	17	20	24	27	31
0.6	0.7257	0.7291	0.7324	0.7357	0.7389	0.7422	0.7454	0.7486	0.7517	0.7549	3	7	10	13	16	19	23	26	29
0.7	0.7580	0.7611	0.7642	0.7673	0.7704	0.7734	0.7764	0.7794	0.7823	0.7852	3	6	9	12	15	18	21	24	27
0.8	0.7881	0.7910	0.7939	0.7967	0.7995	0.8023	0.8051	0.8078	0.8106	0.8133	3	5	8	11	14	16	19	22	25
2.5	0.9938	0.9940	0.9941	0.9943	0.9945	0.9946	0.9948	0.9949	0.9951	0.9952	0	0	0	1	1	1	1	1	1
2.6	0.9953	0.9955	0.9956	0.9957	0.9959	0.9960	0.9961	0.9962	0.9963	0.9964	0	0	0	0	1	1	1	1	1
2.7	0.9965	0.9966	0.9967	0.9968	0.9969	0.9970	0.9971	0.9972	0.9973	0.9974	0	0	0	0	0	1	1	1	1
2.8	0.9974	0.9975	0.9976	0.9977	0.9977	0.9978	0.9979	0.9979	0.9980	0.9981	0	0	0	0	0	0	0	1	1
2.9	0.9981	0.9982	0.9982	0.9983	0.9984	0.9984	0.9985	0.9985	0.9986	0.9986	0	0	0	0	0	0	0	0	0

$\Phi(2.9) = 0.9981$

▲ **Figure 7.2** Extract from tables of $\Phi(z)$

This gives $\Phi(2.9) = 0.9981$, and so $1 - \Phi(2.9) = 0.0019$.

The probability of a randomly selected adult male being 194.3 cm or over is 0.0019. One man in slightly more than 500 is at least as tall as UK Beanpole.

7.1 Using normal distribution tables

The function $\Phi(z)$ gives the area under the normal distribution curve to the *left* of the value z, that is the shaded area in Figure 7.3 (it is the cumulative distribution function). The total area under the curve is 1, and the area given by $\Phi(z)$ represents the probability of a value smaller than z.

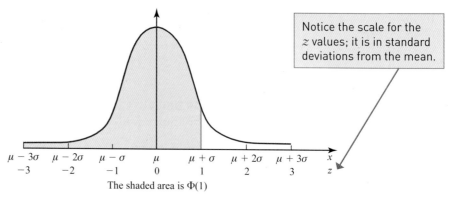

Notice the scale for the z values; it is in standard deviations from the mean.

The shaded area is $\Phi(1)$

▲ **Figure 7.3**

If the variable X has mean μ and standard deviation σ then x, a particular value of X, is transformed into z by the equation

$$z = \frac{x - \mu}{\sigma}$$

z is a particular value of the variable Z which has mean 0 and standard deviation 1 and is the *standardised* form of the normal distribution.

	Actual distribution, X	Standardised distribution, Z
Mean	μ	0
Standard deviation	σ	1
Particular value	x	$z = \dfrac{x - \mu}{\sigma}$

Notice how lower case letters, x and z, are used to indicate particular values of the random variables, whereas upper case letters, X and Z, are used to describe or name those variables.

Normal distribution tables are easy to use but you should always make a point of drawing a diagram and shading the region you are interested in.

It is often helpful to know that in a normal distribution, roughly:

» 68% of the values lie within ±1 standard deviation of the mean

» 95% of the values lie within ±2 standard deviations of the mean

» 99.75% of the values lie within ±3 standard deviations of the mean.

The notation $N(\mu, \sigma^2)$ is used to describe this distribution. The mean, μ, and standard deviation, σ (or variance, σ^2), are the two parameters used to define the distribution. Once you know their values, you know everything there is to know about the distribution. The standardised variable Z has mean 0 and variance 1, so its distribution is $N(0, 1)$.

| Example 7.1 |

Assuming the distribution of the heights of adult men is normal, with mean 174 cm and standard deviation 7 cm, find the probability that a randomly selected adult man is:

(i)   under 185 cm

(ii)   over 185 cm

(iii)   over 180 cm

(iv)   between 180 cm and 185 cm

(v)   under 170 cm

giving your answers to 2 significant figures.

Solution

The mean height, $\mu = 174$ cm.

The standard deviation, $\sigma = 7$ cm.

(i)   The probability that an adult man selected at random is under 185 cm. The area required is that shaded in Figure 7.4.

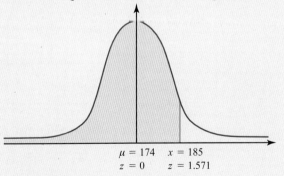

$\mu = 174 \quad x = 185$
$z = 0 \quad\quad z = 1.571$

▲ **Figure 7.4**

$x = 185$ cm

and so $z = \dfrac{185 - 174}{7} = 1.571$ ←

> For calculations that involve standardisation, you should show full details of your working.

Look up the value of $\Phi(z)$ in a normal distribution table.   ➡

z	0	1	2	3	4	5	6	7	8	9	1	2	3	4	5	6	7	8	9
															ADD				
0.0	0.5000	0.5040	0.5080	0.5120	0.5160	0.5199	0.5239	0.5279	0.5319	0.5359	4	8	12	16	20	24	28	32	36
0.1	0.5398	0.5438	0.5478	0.5517	0.5557	0.5596	0.5636	0.5675	0.5714	0.5753	4	8	12	16	20	24	28	32	36
1.4	0.9192	0.9207	0.9222	0.9236	0.9251	0.9265	0.9279	0.9292	0.9306	0.9319	1	3	4	6	7	8	10	11	13
1.5	0.9332	0.9345	0.9357	0.9370	0.9382	0.9394	0.9406	0.9418	0.9429	0.9441	1	2	4	5	6	7	8	10	11
1.6	0.9452	0.9463	0.9474	0.9484	0.9495	0.9505	0.9515	0.9525	0.9535	0.9545	1	2	3	4	5	6	7	8	9
1.7	0.9554	0.9564	0.9573	0.9582	0.9591	0.9599	0.9608	0.9616	0.9625	0.9633	1	2	3	4	4	5	6	7	8
1.8	0.9641	0.9649	0.9656	0.9664	0.9671	0.9678	0.9686	0.9693	0.9699	0.9706	1	1	2	3	4	4	5	6	6

▲ **Figure 7.5** Extract from tables of $\Phi(z)$

$$\Phi(1.571) = 0.9418 + 0.0001$$

$$= 0.9419$$

$$= 0.94 \qquad (2 \text{ s.f.})$$

Answer: The probability that an adult man selected at random is under 185 cm is 0.94.

(ii)   The probability that an adult man selected at random is over 185 cm.
The area required is the complement of that for part (i) (see Figure 7.6).

$$\text{Probability} = 1 - \Phi(1.571)$$

$$= 1 - 0.9419$$

$$= 0.0581$$

$$= 0.058 \qquad (2 \text{ s.f.})$$

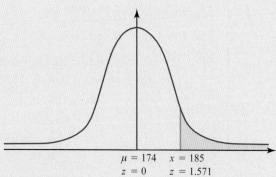

▲ **Figure 7.6**

Answer: The probability that an adult man selected at random is over 185 cm is 0.058.

(iii)   The probability that an adult man selected at random is over 180 cm.

$$x = 180 \text{ cm} \quad \text{and so} \quad z = \frac{180 - 174}{7} = 0.857$$

The area required $= 1 - \Phi(0.857)$
$= 1 - 0.8042$
$= 0.1958$
$= 0.20 \quad (2 \text{ s.f.})$

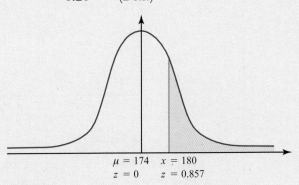

▲ **Figure 7.7**

Answer: The probability that an adult man selected at random is over 180 cm is 0.20.

(iv)   The probability that an adult man selected at random is between 180 cm and 185 cm.

The required area is shown in Figure 7.8. It is

$$\Phi(1.571) - \Phi(0.857) = 0.9419 - 0.8042$$
$$= 0.1377$$
$$= 0.14 \quad (2 \text{ s.f.})$$

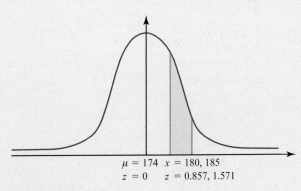

▲ **Figure 7.8**

Answer: The probability that an adult man selected at random is over 180 cm but under 185 cm is 0.14.

(v) The probability that an adult man selected at random is under 170 cm.

In this case $\quad x = 170$

and so $\quad z = \dfrac{170 - 174}{7} = -0.571$

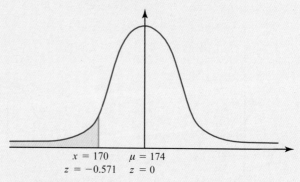

$$x = 170 \qquad \mu = 174$$
$$z = -0.571 \qquad z = 0$$

▲ **Figure 7.9**

However, when you come to look up $\Phi(-0.571)$, you will find that only positive values of z are given in your tables. You overcome this problem by using the symmetry of the normal curve. The area you want in this case is that to the left of -0.571 and this is clearly just the same as that to the right of $+0.571$ (see Figure 7.10).

So $\qquad \Phi(-0.571) = 1 - \Phi(0.571)$

$$= 1 - 0.716$$

$$= 0.284$$

$$= 0.28 \ (2 \text{ s.f.})$$

Answer: The probability that an adult man selected at random is under 170 cm is 0.28.

Note:

These graphs illustrate that $\Phi(-z) = 1 - \Phi(z)$.

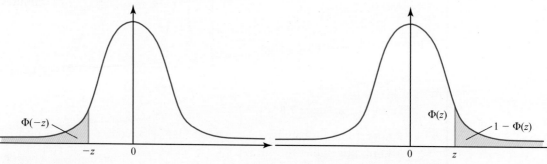

▲ **Figure 7.10**

7.2 The normal curve

All normal curves have the same basic shape, so that by scaling the two axes suitably you can always fit one normal curve exactly on top of another one.

The curve for the normal distribution with mean μ and standard deviation σ (i.e. variance σ^2) is given by the function $\phi(x)$ in

$$\phi(x) = \frac{1}{\sigma\sqrt{2\pi}}e^{-\frac{1}{2}\left(\frac{x-\mu}{\sigma}\right)^2}$$

After the variable X has been transformed to Z using $z = \dfrac{x-\mu}{\sigma}$ the form of the curve (now standardised) becomes

$$\phi(z) = \frac{1}{\sqrt{2\pi}}e^{-\frac{1}{2}z^2}$$

However, the exact shape of the normal curve is often less useful than the area underneath it, which represents a probability. For example, the probability that $Z \leqslant 2$ is given by the shaded area in Figure 7.11.

Easy though it looks, the function $\phi(z)$ cannot be integrated algebraically to find the area under the curve; this can only be found by using a numerical method. The values found by doing so are given as a table, and this area function is called $\Phi(z)$.

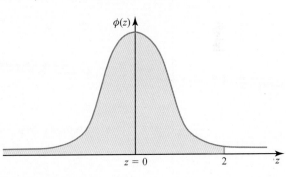

▲ **Figure 7.11**

Example 7.2

Skilled operators make a particular component for an engine. The company believes that the time taken to make this component may be modelled by the normal distribution with mean 95 minutes and standard deviation 4 minutes.

Assuming the company's belief to be true, find the probability that the time taken to make one of these components, selected at random, was:

(i)   over 97 minutes

(ii)   under 90 minutes

(iii)   between 90 and 97 minutes.

Sheila believes that the company is allowing too long for the job and invites them to time her. They find that only 10% of the components take her over 90 minutes to make, and that 20% take her less than 70 minutes.

(iv)   Estimate the mean and standard deviation of the time Sheila takes.   ➔

Solution

According to the company $\mu = 95$ and $\sigma = 4$ so the distribution is $N(95, 4^2)$.

(i)   The probability that a component required over 97 minutes.

$$z = \frac{97 - 95}{4} = 0.5$$

The probability is represented by the shaded area in Figure 7.12 and is given by

$$1 - \Phi(0.5) = 1 - 0.6915$$
$$= 0.3085$$
$$= 0.309 \ (3 \ \text{s.f.})$$

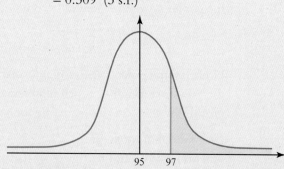

▲ Figure 7.12

Answer: The probability it took the operator over 97 minutes to manufacture a randomly selected component is 0.309.

(ii)   The probability that a component required under 90 minutes.

$$z = \frac{90 - 95}{4} = -1.25$$

The probability is represented by the shaded area in Figure 7.13 and given by

$$1 - \Phi(1.25) = 1 - 0.8944$$
$$= 0.1056$$
$$= 0.106 \ (3 \ \text{s.f.})$$

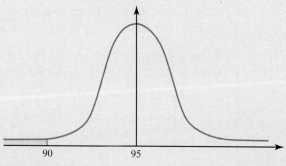

▲ Figure 7.13

Answer: The probability it took the operator under 90 minutes to manufacture a randomly selected component is 0.106.

(iii) The probability that a component required between 90 and 97 minutes.

The probability is represented by the shaded area in Figure 7.14 and given by

$$1 - 0.1056 - 0.3085 = 0.5859$$

$$= 0.586 \quad (3 \text{ s.f.})$$

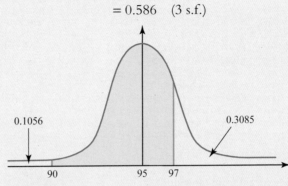

0.1056   0.3085

90   95   97

▲ **Figure 7.14**

Answer: The probability it took the operator between 90 and 97 minutes to manufacture a randomly selected component is 0.586.

(iv) Estimate the mean and standard deviation of the time Sheila takes.

The question has now been put the other way round. You have to infer the mean, μ, and standard deviation, σ, from the areas under different parts of the graph.

10% take her 90 minutes or more. This means that the shaded area in Figure 7.15 is 0.1.

$$z = \frac{90 - \mu}{\sigma}$$

$$\Phi(z) = 1 - 0.1 = 0.9$$

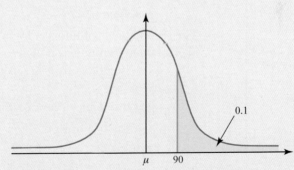

0.1

μ 90

▲ **Figure 7.15**

You now use the table of $\Phi(z) = p$ in reverse. $z = 1.28$ has a probability of 0.8997 which is as close to 0.9 as you can get using this middle part of the table. However, you can achieve greater accuracy by looking at the right-hand columns as well: $z = 1.281$ has a probability of 0.8999 and $z = 1.282$ has a probability of 0.9001. So the best value for z is 1.2815.

z	0	1	2	3	4	5	6	7	8	9	1	2	3	4	5	6	7	8	9
															ADD				
0.0	0.5000	0.5040	0.5080	0.5120	0.5160	0.5199	0.5239	0.5279	0.5319	0.5359	4	8	12	16	20	24	28	32	36
0.1	0.5398	0.5438	0.5478	0.5517	0.5557	0.5596	0.5636	0.5675	0.5714	0.5753	4	8	12	16	20	24	28	32	36
0.2	0.5793	0.5832	0.5871	0.5910	0.5948	0.5987	0.6026	0.6064	0.6103	0.6141	4	8	12	15	19	23	27	31	35
0.3	0.6179	0.6217	0.6255	0.6293	0.6331	0.6368	0.6406	0.6443	0.6480	0.6517	4	7	11	15	19	22	26	30	34
0.4	0.6554	0.6591	0.6628	0.6664	0.6700	0.6736	0.6772	0.6808	0.6844	0.6879	4	7	11	14	18	22	25	29	32
0.5	0.6915	0.6950	0.6985	0.7019	0.7054	0.7088	0.7123	0.7157	0.7190	0.7224	3	7	10	14	17	20	24	27	31
0.6	0.7257	0.7291	0.7324	0.7357	0.7389	0.7422	0.7454	0.7486	0.7517	0.7549	3	7	10	13	16	19	23	26	29
0.7	0.7580	0.7611	0.7642	0.7673	0.7704	0.7734	0.7764	0.7794	0.7823	0.7852	3	6	9	12	15	18	21	24	27
0.8	0.7881	0.7910	0.7939	0.7967	0.7995	0.8023	0.8051	0.8078	0.8106	0.8133	3	5	8	11	14	16	19	22	25
0.9	0.8159	0.8186	0.8212	0.8238	0.8264	0.8289	0.8315	0.8340	0.8365	0.8389	3	5	8	10	13	15	18	20	23
1.0	0.8413	0.8438	0.8461	0.8485	0.8508	0.8531	0.8554	0.8577	0.8599	0.8621	2	5	7	9	12	14	16	19	21
1.1	0.8643	0.8665	0.8686	0.8708	0.8729	0.8749	0.8770	0.8790	0.8810	0.8830	2	4	6	8	10	12	14	16	18
1.2	0.8849	0.8869	0.8888	0.8907	0.8925	0.8944	0.8962	0.8980	0.8997	0.9015	2	4	6	7	9	11	13	15	17
1.3	0.9032	0.9049	0.9066	0.9082	0.9099	0.9115	0.9131	0.9147	0.9162	0.9177	2	3	5	6	8	10	11	13	14
1.4	0.9192	0.9207	0.9222	0.9236	0.9251	0.9265	0.9279	0.9292	0.9306	0.9319	1	3	4	6	7	8	10	11	13

$\Phi^{-1}(0.9) = 1.2815$

▲ **Figure 7.16** **Extract from tables of $\Phi(z)$**

Returning to the problem, you now know that

$$\frac{90 - \mu}{\sigma} = 1.2815 \quad \Rightarrow \quad 90 - \mu = 1.2815\sigma. \qquad \text{①}$$

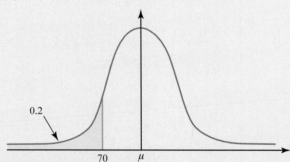

▲ **Figure 7.17**

The second piece of information, that 20% of components took Sheila under 70 minutes, is illustrated in Figure 7.17.

$$z = \frac{70 - \mu}{\sigma}$$

(z has a negative value in this case, the point being to the left of the mean.)

$$\Phi(z) = 0.2$$

and so, by symmetry,

$$\Phi(-z) = 1 - 0.2 = 0.8.$$

Using the table of the normal function gives

$$-z = 0.842 \quad \text{or} \quad z = -0.842.$$

This gives a second equation for μ and σ.

$$\frac{70 - \mu}{\sigma} = -0.842 \quad \Rightarrow \quad 70 - \mu = -0.842\sigma. \qquad \text{②}$$

You now solve equations ① and ② simultaneously.

$$90 - \mu = 1.2815\sigma \qquad ①$$
$$70 - \mu = -0.842\sigma \qquad ②$$

Subtract $\quad 20 \quad = 2.1235\sigma$

$$\sigma = 9.418 = 9.42 \quad \text{(3 s.f.)}$$

and $\qquad\qquad \mu = 77.930 = 77.9 \quad \text{(3 s.f.)}$

Answer: Sheila's mean time is 77.9 minutes with standard deviation 9.42 minutes.

1 The distribution of the heights of some plants is normal and has a mean of 40 cm and a standard deviation of 2 cm. Find the probability that a randomly selected plant is:

(i)   under 42 cm

(ii)   over 42 cm

(iii)   over 40 cm

(iv)   between 40 and 42 cm.

2 The distribution of the masses of some baby parrots is normal and has a mean of 60 g and a standard deviation of 5 g. Find the probability that a randomly selected bird is:

(i)   under 63 g

(ii)   over 63 g

(iii)   over 68 g

(iv)   between 63 and 68 g.

3 The distribution of the mass of sweets in a bag is normal and has a mean of 100 g and a standard deviation 2 g. Find the probability that a randomly selected bag is:

(i)   under 98 g

(ii)   over 98 g

(iii)   under 102 g

(iv)   between 98 and 102 g.

4 The distribution of the heights of 18-year-old girls may be modelled by the normal distribution with mean 162.5 cm and standard deviation 6 cm. Find the probability that the height of a randomly selected 18-year-old girl is:

(i)   under 168.5 cm

(ii)   over 174.5 cm

(iii)   between 168.5 and 174.5 cm.

7

M **5** A pet shop has a tank of goldfish for sale. All the fish in the tank were hatched at the same time and their weights may be taken to be normally distributed with mean 100 g and standard deviation 10 g. Melanie is buying a goldfish and is invited to catch the one she wants in a small net. In fact the fish are much too quick for her to be able to catch any particular fish, and the one which she eventually nets is selected at random. Find the probability that its weight is:

(i) over 115 g

(ii) under 105 g

(iii) between 105 and 115 g.

M **6** When he makes instant coffee, Tony puts a spoonful of powder into a mug. The weight of coffee in grams on the spoon may be modelled by the normal distribution with mean 5 g and standard deviation 1 g. If he uses more than 6.5 g Julia complains that it is too strong and if he uses less than 4 g she tells him it is too weak. Find the probability that he makes the coffee:

(i) too strong

(ii) too weak

(iii) alright.

CP **7** A biologist finds a nesting colony of a previously unknown sea bird on a remote island. She is able to take measurements on 100 of the eggs before replacing them in their nests. She records their weights, w g, in this frequency table.

Weight, w	$25 < w \leqslant 27$	$27 < w \leqslant 29$	$29 < w \leqslant 31$	$31 < w \leqslant 33$	$33 < w \leqslant 35$	$35 < w \leqslant 37$
Frequency	2	13	35	33	17	0

(i) Find the mean and standard deviation of these data.

(ii) Assuming the weights of the eggs for this type of bird are normally distributed and that their mean and standard deviation are the same as those of this sample, find how many eggs you would expect to be in each of these categories.

(iii) Do you think the assumption that the weights of the eggs are normally distributed is reasonable?

PS **8** A machine is set to produce nails of length 10 cm, with standard deviation 0.05 cm. The lengths of the nails are normally distributed.

(i) Find the percentage of nails produced between 9.95 cm and 10.08 cm in length.

The machine's setting is moved by a careless apprentice with the consequence that 16% of the nails are under 5.2 cm in length and 20% are over 5.3 cm.

(ii) Find the new mean and standard deviation.

9   The random variable X has the distribution $N(\mu, \sigma^2)$. It is given that $P(X < 54.1) = 0.5$ and $P(X > 50.9) = 0.8665$. Find the values of μ and σ.

Cambridge International AS & A Level Mathematics
9709 Paper 61 Q2 November 2015

10 The petrol consumption of a certain type of car has a normal distribution with mean 24 kilometres per litre and standard deviation 4.7 kilometres per litre. Find the probability that the petrol consumption of a randomly chosen car of this type is between 21.6 kilometres per litre and 28.7 kilometres per litre.

Cambridge International AS & A Level Mathematics
9709 Paper 61 Q1 June 2014

11 Lengths of a certain type of white radish are normally distributed with mean μ cm and standard deviation σ cm. 4% of these radishes are longer than 12 cm and 32% are longer than 9 cm. Find μ and σ.

Cambridge International AS & A Level Mathematics
9709 Paper 61 Q2 June 2014

12 A farmer finds that the weights of sheep on his farm have a normal distribution with mean 66.4 kg and standard deviation 5.6 kg.

(i)   250 sheep are chosen at random. Estimate the number of sheep which have a weight of between 70 kg and 72.5 kg.

(ii) The proportion of sheep weighing less than 59.2 kg is equal to the proportion weighing more than y kg. Find the value of y.

Another farmer finds that the weights of sheep on his farm have a normal distribution with mean μ kg and standard deviation 4.92 kg. 25% of these sheep weigh more than 67.5 kg.

(iii) Find the value of μ.

Cambridge International AS & A Level Mathematics
9709 Paper 61 Q6 November 2014

13 It is given that $X \sim N(30, 49)$, $Y \sim N(30, 16)$ and $Z \sim N(50, 16)$. On a single diagram, with the horizontal axis going from 0 to 70, sketch three curves to represent the distributions of X, Y and Z.

Cambridge International AS & A Level Mathematics
9709 Paper 61 Q1 November 2013

14 Lengths of a certain type of carrot have a normal distribution with mean 14.2 cm and standard deviation 3.6 cm.

(i)   8% of carrots are shorter than c cm. Find the value of c.

(ii) Rebekah picks 7 carrots at random. Find the probability that at least 2 of them have lengths between 15 cm and 16 cm.

Cambridge International AS & A Level Mathematics
9709 Paper 61 Q5 November 2013

7

7.2 The normal curve

15 Tyre pressures on a certain type of car independently follow a normal distribution with mean 1.9 bars and standard deviation 0.15 bars.

(i) Find the probability that all four tyres on a car of this type have pressures between 1.82 bars and 1.92 bars.

(ii) Safety regulations state that the pressures must be between $1.9 - b$ bars and $1.9 + b$ bars. It is known that 80% of tyres are within these safety limits. Find the safety limits.

Cambridge International AS & A Level Mathematics
9709 Paper 6 Q6 June 2005

16 The lengths of fish of a certain type have a normal distribution with mean 38 cm. It is found that 5% of the fish are longer than 50 cm.

(i) Find the standard deviation.

(ii) When fish are chosen for sale, those shorter than 30 cm are rejected. Find the proportion of fish rejected.

(iii) 9 fish are chosen at random. Find the probability that at least one of them is longer than 50 cm.

Cambridge International AS & A Level Mathematics
9709 Paper 6 Q3 June 2006

17 (a) The random variable X is normally distributed. The mean is twice the standard deviation. It is given that $P(X > 5.2) = 0.9$. Find the standard deviation.

(b) A normal distribution has mean μ and standard deviation σ. If 800 observations are taken from this distribution, how many would you expect to be between $\mu - \sigma$ and $\mu + \sigma$?

Cambridge International AS & A Level Mathematics
9709 Paper 6 Q3 June 2007

18 In a certain country the time taken for a common infection to clear up is normally distributed with mean μ days and standard deviation 2.6 days. 25% of these infections clear up in less than 7 days.

(i) Find the value of μ.

In another country the standard deviation of the time taken for the infection to clear up is the same as in part (i) but the mean is 6.5 days. The time taken is normally distributed.

(ii) Find the probability that, in a randomly chosen case from this country, the infection takes longer than 6.2 days to clear up.

Cambridge International AS & A Level Mathematics
9709 Paper 6 Q4 June 2008

19

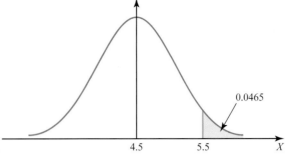

The random variable X has a normal distribution with mean 4.5. It is given that $P(X > 5.5) = 0.0465$ (see diagram).

(i) Find the standard deviation of X.

(ii) Find the probability that a random observation of X lies between 3.8 and 4.8.

Cambridge International AS & A Level Mathematics
9709 Paper 6 Q4 November 2007

20 (i) The daily minimum temperature in degrees Celsius (°C) in January in Ottawa is a random variable with distribution N(−15.1, 62.0). Find the probability that a randomly chosen day in January in Ottawa has a minimum temperature above 0 °C.

(ii) In another city the daily minimum temperature in °C in January is a random variable with distribution N(μ, 40.0). In this city the probability that a randomly chosen day in January has a minimum temperature above 0 °C is 0.8888. Find the value of μ.

Cambridge International AS & A Level Mathematics
9709 Paper 6 Q3 November 2008

21 The times for a certain car journey have a normal distribution with mean 100 minutes and standard deviation 7 minutes. Journey times are classified as follows:

'short'   (the shortest 33% of times)

'long'   (the longest 33% of times)

'standard'   (the remaining 34% of times).

(i) Find the probability that a randomly chosen car journey takes between 85 and 100 minutes.

(ii) Find the least and greatest times for 'standard' journeys.

Cambridge International AS & A Level Mathematics
9709 Paper 61 Q3 November 2009

7.3 Modelling discrete situations

Although the normal distribution applies strictly to a continuous variable, it is also common to use it in situations where the variable is discrete providing that:

➤➤ the distribution is approximately normal; this requires that the steps in its possible values are small compared with its standard deviation

➤➤ **continuity corrections** are applied where appropriate.

The meaning of the term 'continuity correction' is explained in the following example.

Example 7.3

The result of an Intelligence Quotient (IQ) test is an integer score, X. Tests are designed so that X has a mean value of 100 with standard deviation 15. A large number of people have their IQs tested. What proportion of them would you expect to have IQs measuring between 106 and 110 (inclusive)?

Solution

Although the random variable X is an integer and hence discrete, the steps of 1 in its possible values are small compared with the standard deviation of 15. So it is reasonable to treat it as if it is continuous.

If you assume that an IQ test is measuring innate, natural intelligence (rather than the results of learning), then it is reasonable to assume a normal distribution.

If you draw the probability distribution function for the discrete variable X it looks like Figure 7.18. The area you require is the total of the five bars representing 106, 107, 108, 109 and 110.

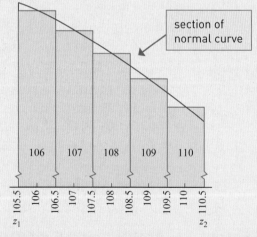

▲ **Figure 7.18**

The equivalent section of the normal curve would run not from 106 to 110 but from 105.5 to 110.5, as you can see in Figure 7.18. When you change from the discrete scale to the continuous scale, the numbers 106, 107, etc. no longer represent the whole intervals, just their centre points.

So the area you require under the normal curve is given by $\Phi(z_2) - \Phi(z_1)$

where $z_1 = \dfrac{105.5 - 100}{15}$ and $z_2 = \dfrac{110.5 - 100}{15}$.

This is

$$\Phi(0.7000) - \Phi(0.3667) = 0.7580 - 0.6431 = 0.1149$$

Answer: The proportion of IQs between 106 and 110 (inclusive) should be approximately 11%.

In this calculation, both end values needed to be adjusted to allow for the fact that a continuous distribution was being used to approximate a discrete one. These adjustments, $106 \rightarrow 105.5$ and $110 \rightarrow 110.5$, are called continuity corrections. Whenever a discrete distribution is approximated by a continuous one a continuity correction may need to be used.

You must always think carefully when applying a continuity correction. Should the corrections be added or subtracted? In this case 106 and 110 are inside the required area and so any value (like 105.7 or 110.4) which would round to them must be included. It is often helpful to draw a sketch to illustrate the region you want, like the one in Figure 7.18.

If the region of interest is given in terms of inequalities, you should look carefully to see whether they are inclusive ($\leqslant$ or $\geqslant$) or exclusive ($<$ or $>$). For example $20 \leqslant X \leqslant 30$ becomes $19.5 \leqslant X < 30.5$ whereas $20 < X < 30$ becomes $20.5 \leqslant X < 29.5$.

Two particularly common situations are when the normal distribution is used to approximate the binomial and the Poisson distributions. (You will learn about the Poisson distribution if you study *Probability & Statistics 2*.)

7.4 Using the normal distribution as an approximation for the binomial distribution

You may use the normal distribution as an approximation for the binomial, $B(n, p)$ (where n is the number of trials each having probability p of success) when:

» n is large

» p is not too close to 0 or 1.

A rough way of judging whether n is large enough is to require that both $np > 5$ and $nq > 5$, where $q = 1 - p$.

These conditions ensure that the distribution is reasonably symmetrical and not skewed away from either end, see Figure 7.19.

The parameters for the normal distribution are then

$$\text{Mean:} \quad \mu = np \qquad \text{Variance:} \quad \sigma^2 = npq = np(1-p)$$

so that it can be denoted by $N(np, npq)$.

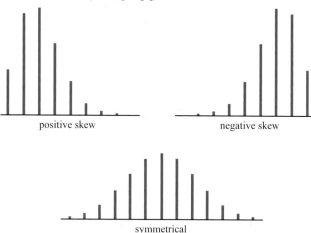

positive skew

negative skew

symmetrical

▲ **Figure 7.19**

Example 7.4

This is a true story. During voting at an election, an exit poll of 1700 voters indicated that 50% of people had voted for a particular candidate. When the votes were counted it was found that he had in fact received 57% support.

850 of the 1700 people interviewed said they had voted for the candidate but 57% of 1700 is 969, a much higher number. What went wrong? Is it possible to be so far out just by being unlucky and asking the wrong people?

Solution

The situation of selecting a sample of 1700 people and asking them if they voted for one candidate or not is one that is modelled by the binomial distribution, in this case $B(1700, 0.57)$.

In theory you could multiply out $(0.43 + 0.57t)^{1700}$ and use that to find the probabilities of getting 0, 1, 2, ..., 850 supporters of this candidate in your sample of 1700. In practice such a method would be impractical because of the work involved.

What you can do is to use a normal approximation. The required conditions are fulfilled: at 1700, n is certainly not small; $p = 0.57$ is near neither 0 nor 1.

The parameters for the normal approximation are given by

$$\mu = np = 1700 \times 0.57 = 969$$

$$\sigma = \sqrt{npq} = \sqrt{1700 \times 0.57 \times 0.43} = 20.4$$

You will see that the standard deviation, 20.4, is large compared with the steps of 1 in the number of supporters of this candidate.

The probability of getting no more than 850 supporters of this candidate, $P(X \leqslant 850)$, is given by $\Phi(z)$, where

$$z = \frac{850.5 - 969}{20.4} = -5.8$$

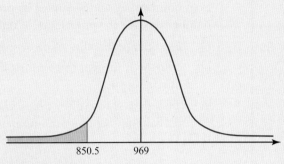

▲ **Figure 7.20**

(Notice the continuity correction making 850 into 850.5.)

This is beyond the range of most tables and corresponds to a probability of about 0.000 01. The probability of a result as extreme as this is thus 0.000 02 (allowing for an equivalent result in the tail above the mean). It is clearly so unlikely that this was a result of random sampling that another explanation must be found.

❯ What do you think went wrong with the exit poll? Remember this really did happen.

Exercise 7B

1   25% of Flapper Fish have red spots, the rest have blue spots. A fisherman nets 10 Flapper Fish. What are the probabilities that:

(i)   exactly 8 have blue spots

(ii)   at least 8 have blue spots?

A large number of samples, each of 100 Flapper Fish, are taken.

(iii)   What is the mean and the standard deviation of the number of red-spotted fish per sample?

(iv)   What is the probability of a sample of 100 Flapper Fish containing over 30 with red spots?

2 Assume that, for a randomly chosen person, their next birthday is equally likely to occur on any day of the week, independently of any other person's birthday. Find the probability that, out of 350 randomly chosen people, at least 47 will have their next birthday on a Monday.

Cambridge International AS & A Level Mathematics
9709 Paper 61 Q2 June 2013

3 The faces of a biased die are numbered 1, 2, 3, 4, 5 and 6. The probabilities of throwing odd numbers are all the same. The probabilities of throwing even numbers are all the same. The probability of throwing an odd number is twice the probability of throwing an even number.

(i) Find the probability of throwing a 3.

(ii) The die is thrown three times. Find the probability of throwing two 5s and one 4.

(iii) The die is thrown 100 times. Use an approximation to find the probability that an even number is thrown at most 37 times.

Cambridge International AS & A Level Mathematics
9709 Paper 61 Q7 November 2015

4 It is known that, on average, 2 people in 5 in a certain country are overweight. A random sample of 400 people is chosen. Using a suitable approximation, find the probability that fewer than 165 people in the sample are overweight.

Cambridge International AS & A Level Mathematics
9709 Paper 6 Q1 June 2005

5 A survey of adults in a certain large town found that 76% of people wore a watch on their left wrist, 15% wore a watch on their right wrist and 9% did not wear a watch.

(i) A random sample of 14 adults was taken. Find the probability that more than 2 adults did not wear a watch.

(ii) A random sample of 200 adults was taken. Using a suitable approximation, find the probability that more than 155 wore a watch on their left wrist.

Cambridge International AS & A Level Mathematics
9709 Paper 6 Q7 June 2006

6 On a certain road 20% of the vehicles are trucks, 16% are buses and the remainder are cars.

(i) A random sample of 11 vehicles is taken. Find the probability that fewer than 3 are buses.

(ii) A random sample of 125 vehicles is now taken. Using a suitable approximation, find the probability that more than 73 are cars.

Cambridge International AS & A Level Mathematics
9709 Paper 6 Q3 June 2009

7 On any occasion when a particular gymnast performs a certain routine, the probability that she will perform it correctly is 0.65, independently of all other occasions.

(i) Find the probability that she will perform the routine correctly on exactly 5 occasions out of 7.

(ii) On one day she performs the routine 50 times. Use a suitable approximation to estimate the probability that she will perform the routine correctly on fewer than 29 occasions.

(iii) On another day she performs the routine n times. Find the smallest value of n for which the expected number of correct performances is at least 8.

Cambridge International AS & A Level Mathematics
9709 Paper 6 Q6 November 2007

8 In the holidays Martin spends 25% of the day playing computer games. Martin's friend phones him once a day at a randomly chosen time.

(i) Find the probability that, in one holiday period of 8 days, there are exactly 2 days on which Martin is playing computer games when his friend phones.

(ii) Another holiday period lasts for 12 days. State with a reason whether it is appropriate to use a normal approximation to find the probability that there are fewer than 7 days on which Martin is playing computer games when his friend phones.

(iii) Find the probability that there are at least 13 days of a 40-day holiday period on which Martin is playing computer games when his friend phones.

Cambridge International AS & A Level Mathematics
9709, Paper 61 Q5 June 2010

KEY POINTS

1 The normal distribution with mean μ and standard deviation σ is denoted by $N(\mu, \sigma^2)$.

2 This may be given in standardised form by using the transformation

$$z = \frac{x - \mu}{\sigma}$$

3 In the standardised form, $N(0, 1)$, the mean is 0, and the standard deviation and variance are both 1.

4 The standard normal curve is given by

$$\Phi(z) = \frac{1}{\sqrt{2\pi}} e^{-\frac{1}{2}z^2}$$

5 The area to the left of the value z in the diagram below, representing the probability of a value less than z, is denoted by $\Phi(z)$ and is read from tables.

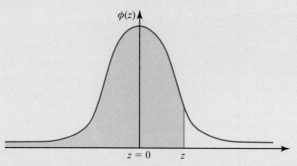

6 The normal distribution may be used to approximate suitable discrete distributions but continuity corrections are then required.

7 The binomial distribution $B(n, p)$ may be approximated by $N(np, npq)$, provided n is large and p is not close to 0 or 1, so that $np > 5$ and $nq > 5$.

LEARNING OUTCOMES

Now that you have finished this chapter, you should be able to

- use the normal distribution as a model

- know the shape of a normal curve and the location of its line of symmetry

- be able to standardise a normal variable

- find probabilities from a normal distribution

- understand how and why a continuity correction is applied when the normal distribution is used as an approximation for the binomial distribution.

Answers

The questions, with the exception of those from past question papers, and all example answers that appear in this book were written by the authors. Cambridge Assessment International Education bears no responsibility for the example answers to questions taken from its past question papers which are contained in this publication.

Non-exact numerical answers should be given correct to three significant figures (or one decimal place for angles in degrees) unless a different level of accuracy is specified in the question. You should avoid rounding figures until reaching your final answer.

Chapter 1

❷ (Page 1)

See text that follows.

❷ (Page 2)

The editor has explained clearly why the investigation is worth doing: there is growing concern about cycling accidents involving children. Good quality data is data that best represents the research topic: in this case it is to establish whether or not the number of accidents is significant.

❷ (Page 2)

The reporter is focusing on two aspects of the investigation: he is looking at cycling accidents in the area over a period of time and he is considering the distribution of ages of accident victims. Both of these sources are relevant to the investigation.

Another thing he might consider is to investigate accidents in a similar community in order to be able to make comparisons

❷ (Page 6)

Not all the branches have leaves. However, all the branches must be shown in order to show correctly the shape of the distribution.

❷ (Page 8)

If the basic stem-and-leaf diagram has too many lines, you may *squeeze* it as shown to the right. In doing this you lose some of the information but you should get a better idea of the shape of the distribution.

Unsqueezed

```
30 | 2 6                    30 | 2 represents 3.02
31 | 4
32 | 0 5
33 | 3
34 | 3 6 7
35 | 0 3 4 4 8
36 | 0 0 4 4 4
37 | 0 1 1 3 3 4 8
38 | 3 3 3 5
39 | 0 0 4
40 | 2
41 | 0 0 1 1 4 4
42 |
43 | 0 2 4
```

Squeezed

The data is first rounded to one decimal place, so 3.02 becomes 3.0 etc.

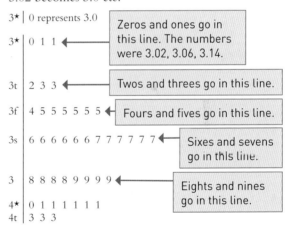

```
3* | 0          represents 3.0
3* | 0 1 1      Zeros and ones go in this line. The numbers were 3.02, 3.06, 3.14.
3t | 2 3 3      Twos and threes go in this line.
3f | 4 5 5 5 5 5 5   Fours and fives go in this line.
3s | 6 6 6 6 6 6 7 7 7 7 7 7   Sixes and sevens go in this line.
3  | 8 8 8 8 9 9 9 9   Eights and nines go in this line.
4* | 0 1 1 1 1 1 1
4t | 3 3 3
```

❷ (Page 8)

Positive and negative data can be represented on a stem-and-leaf diagram in the following way.
Data set:
−36 −32 −28 −25 −24 −20 −18 −15
−12 −6   5   8   12   13   18   26

$n = 16$

$-3 \mid 2$ represents -32

-3	6 2
-2	8 5 4 0
-1	8 5 2
-0	6
0	5 8
1	2 3 8
2	6

Exercise 1A (Page 9)

1 3.27 cm, 3.32 cm, 3.36 cm, 3.43 cm, 3.45 cm, 3.49 cm, 3.50 cm, 3.52 cm, 3.56 cm, 3.56 cm, 3.58 cm, 3.61 cm, 3.61 cm, 3.64 cm, 3.72 cm

2 0.083 mm, 0.086 mm, 0.087 mm, 0.090 mm, 0.091 mm, 0.094 mm, 0.098 mm, 0.102 mm, 0.103 mm, 0.105 mm, 0.108 mm, 0.109 mm, 0.109 mm, 0.110 mm, 0.111 mm, 0.114 mm, 0.123 mm, 0.125 mm, 0.131 mm

3 $n = 13$

$21 \mid 2$ represents 0.212

21	2
22	3 6
23	0 3 7
24	1 2 8
25	3 3 9
26	2

4 $n = 10$

$78 \mid 1$ represents 78.1

78	1
79	4 6
80	4 8
81	3 7 9
82	0 5

5 0.013 m, 0.089 m, 1.79 m, 3.43 m, 3.51 m, 3.57 m, 3.59 m, 3.60 m, 3.64 m, 3.66 m, 3.68 m, 3.71 m, 3.71 m, 3.73 m, 3.78 m, 3.79 m, 3.80 m, 3.85 m, 3.94 m, 7.45 m, 10.87 m

6 (i) 5. It is clearly a mistake as 5-year-olds don't drive.

(ii) $n = 39$

$2 \mid 8$ represents 28 years of age

0	
1	9
2	2 6 6 8 8 9 9
3	2 3 3 4 5 5 5 6 6 7 7 7 8 8 8 9 9
4	1 4 5 5 6
5	2 5 9
6	0 1 2 3 6
7	
8	1

(iii) The distribution has positive skew.

7 (i) 83 years

(ii)

0	
1	6 8 9 9
2	0 1 1 1 2 3 4 5 6 6 7 8 8 9
3	0 0 0 1 2 2 3 5 9
4	3 3 4 5 5 6 6 7 8 9
5	1 2 2 7
6	
7	
8	3

(iii)

0★	
0	
1★	
1	6 8 9 9
2★	0 1 1 1 2 3 4
2	5 6 6 7 8 8 9
3★	0 0 0 1 2 2 3
3	5 9
4★	3 3 4
4	5 5 6 6 7 8 9
5★	1 2 2
5	7
6★	
6	
7★	
7	
8★	3

(iv) The stem-and-leaf diagram with steps of 10 suggests a slight positive skew. The stretched stem-and-leaf diagram shows a clear bimodal spread to the distribution. The first peak (20s) may indicate first marriages and the second peak (40s) may indicate second marriages.

8 (i) $n = 30$

$2 \mid 1$ represents $21\,°C$

-1	5 4 1
-0	9 7 4 2
0	1 2 3 3 4 7 8 8
1	0 1 4 4 5 8 9 9
2	1 3 6 7 9
3	2 5

(ii) The distribution is approximately symmetrical and unimodal.

9 $n = 87$

$1 \mid 8$ represents 18 marks

1	7 8
2	2 5 6 6 9
3	0 0 2 4 4 5 6 7 7 7 9 9
4	0 0 3 4 4 4 5 5 6 8 9 9
5	0 1 1 1 2 2 3 4 4 4 5 6 7 9
6	0 1 2 4 5 6 6 6 6 7 8 8 9 9
7	0 0 1 3 4 4 5 5 6 6 7
8	0 2 7 7
9	0 0 1 2 4 5 5 6 7 7 8
10	0 0

The distribution is symmetrical apart from a peak in the 90s. There is a large concentration of marks between 30 and 80.

10 (i) $n = 40$

```
1 | 9  represents 19 years of age

1 | 7 9 9 9
2 | 0 1 1 2 2 3 3 3 4 5 6 6 6 6 8 8 8 9
3 | 0 0 1 4 5 7
4 | 0 5
5 | 7 8 8
6 | 5 5 5 6 7 9
7 | 2
```

(ii) The distribution is bimodal. This is possibly because those who hang-glide are the reasonably young and active (average age about 25 years) and those who are retired and have taken it up as a hobby (average age about 60).

11 $n = 50 + 50$

```
                      7 | 5 | 3  means 5.7 kg untreated and
                        |   | 5.3 kg treated
       Untreated                          Treated
              8 8 5 5 | 0 |
7 6 4 4 4 4 2 2 2 1 0 0 | 1 | 8 8 9
     9 9 5 2 2 2 0 0 | 2 | 0 1 3 3 3 4 4 5 5 5 6 7
           3 2 1 0 0 | 3 | 0 0 1 1 2 3 3 3 3 5 5 5 8 8 8
       8 8 5 3 2 1 0 0 | 4 | 1 1 2 2 4 4 4 7 9
       8 6 5 3 2 1 1 0 | 5 | 0 2 3 3 4 4 9
           3 2 1 1 1 | 6 | 1 1 2 3
```

GRO seems to have improved the yield of lime trees, though there is a significant number of untreated trees that are matching the yield of the treated trees.

12 (i) $n = 25$

```
11 | 4  represents 1.14 metres

11 | 4 5 8
12 | 1 4 6 6 8 9 9
13 | 0 0 0 0 0 2 2 3 3
14 | 0 1 2 6
15 | 4
16 | 5
```

(ii) 1.46 m, 1.18 m

(iii) 1.30 m

(iv) The median 1.30 is close to the mean value of this data set (1.32) so the median seems a reasonable estimate of the length.

❷ (Page 17)

The median, as it is not affected by the extreme values.

Exercise 1B (Page 18)

1 (i) mode = 45, mean = 39.6, median = 41

(ii) bimodal, 116 and 132, mean = 122.5, median = 122

(iii) mode = 6, mean = 5.3, median = 6

2 (i) (a) mode = 14 years 8 months, mean = 14 years 5.5 months, median = 14 years 6 months

(b) Small data set so mode is inappropriate. You would expect all the students in one class to be uniformly spread between 14 years 0 months and 15 years, so either of the other measures would be acceptable.

(ii) (a) mode = 0, mean = 53.5, median = 59

(b) The median. Small sample makes the mode unreliable and the mean is influenced by extreme values.

(iii) (a) mode = 0 and 21 (bimodal), mean = 29.4, median = 20

(b) Small sample so the mode is inappropriate; the mean is affected by outliers, so the median is the best choice.

(iv) (a) no unique mode, mean = 3.45, median = 3.5

(b) Anything but the mode will do. The distribution, uniform in theory, means that mean = median. This sample reflects that well.

3 (i)

```
                7 | 13 | 3  means 13.7 minutes
                  |    | for 16-year-olds and
                  |    | 13.3 minutes for
  16-year-olds       9-year-olds

           7 4 | 11 |
           9 8 | 12 |
           7 0 | 13 | 0 2 7
             8 | 14 | 2 4
               | 15 | 0 1 9
             5 | 16 | 0 1 4 7
```

(ii) 15.6 minutes

Exercise 1C (Page 21)

1 (i) mode = 2

(ii) median = 3

(iii) mean = 3.24

2 (i) mode = 39

(ii) median = 40

(iii) mean = 40.3

(iv) The sample has a slight positive skew. Any of the measures would do; however, the mean allows one to calculate the total number of matches.

3 (i) mode = 19
(ii) median = 18
(iii) mean = 17.9
(iv) The outliers affect the mean. As the distribution is, apart from the extremes, reasonably symmetrical, the median or mode are acceptable. The median is the safest for a relatively small data set.

4 (i) mode = 1, mean = 1.35, median = 1
(ii) the median

5 (i) mode = 1, mean = 2.09, median = 2
(ii) the median

❓ (Page 24)

The upper boundaries are not stated. 0– could mean 0–9 or it could mean at least 0.

❓ (Page 26)

The mode can be estimated as in this example for a unimodal histogram.

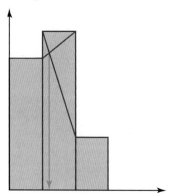

The median can be estimated by interpolation of the interval containing the median or by use of a cumulative frequency curve.
The mean of the original data is 39.73. The estimate is reasonably close at 39.855.

❓ (Page 27)

The median, as it is least affected by extreme values.

❓ (Page 28)

The fairest answer is there is not enough information. Ignoring the journalistic prose, '... our town council rate somewhere between savages and barbarians ...', most of the facts given are correct. However, to say whether or not the council is negligent one would need to compare accident statistics with other *similar* communities. Also, one would need to ask who is responsible for a cyclist's risk of having or not having an accident? Perhaps parents should ensure there is adequate training given to their children, and so on.

❓ (Page 29)

Robert needs to increase his estimate by 0.5 cm (162.32 cm becomes 162.82 cm). The mean of the raw data is 162.86 cm. The estimated value is very close.

Exercise 1D (Page 30)

1 Mid-class values: 114.5, 124.5, 134.5, 144.5, 154.5, 164.5, 174.5, 184.5
Estimated mean = 161 cm (to 3 s.f.)

2 (i) Mid-class values: 24.5, 34.5, 44.5, 54.5, 64.5, 79.5, 104.5
Mean = 48.5 minutes
(ii) The second value seems significantly higher. To make the comparison valid the method of data collection would have to be similar, as would the target children sampled.

3 (i) Mid-interval values: 4.5, 14.5, 24.5, 34.5, 44.5, 54.5, 64.5, 74.5, 84.5
Mean (stated age) = 29.7 years
(ii) Add 0.5; mean age = 30.2 years
(iii) Once adjusted, the estimated mean age compares well with the actual mean of 30.4 years.

4 Mean = 43.1 cm

5 (i) Mid-class values: 25, 75, 125, 175, 250, 400, 750, 3000
Mean = 951 m
(ii) The way in which these data are grouped seems to have a marked effect on the mean. This is probably because the distribution is so skewed.

6 (i) $59.5 \leqslant x < 99.5$
(ii) 138.5 g

❷ (Page 34)

With the item $90 removed the mean is $15.79, compared to $19.50. The extreme value 'dragged' the value of the mean towards it.

❷ (Page 34)

Each deviation is by definition the data value minus the mean. As the mean is *central* to the data, some deviations will be negative (below the mean) and some will be positive (above the mean). The total deviation above the mean cancels out the total deviation below the mean.

❷ (Page 39)

Using 656 instead of the accurate value of 655.71... results in

variance = 430 041.14... − 430 336
$$= -294.85...$$

which, being negative, is impossible.

❷ (Page 39)

With the value 96 omitted, mean = 54.2, standard deviation = 7.9. The value 96 is more than five standard deviations above the new mean value.

Exercise 1E (Page 40)

1 (i) Mean = 2.36

(ii) Standard deviation = 1.49

2 Mean = 6.04, standard deviation = 1.48

3 (i) Mahmood: mean = 1.03, standard deviation = 1.05
Raheem: mean = 1.03, standard deviation = 0.55

(ii) On average they scored the same number of goals but Raheem was more consistent.

4 Mean = 1.1, standard deviation = 1.24

5 Mean = 0.4, standard deviation = 0.4

6 (i) A: mean = 25 °C, standard deviation = 1.41 °C
B: mean = 25 °C, standard deviation = 2.19 °C

(ii) Thermostat A is better. The smaller standard deviation shows it is more consistent in its settings.

(iii) Mean = 24.8 °C, standard deviation = 1.05 °C

7 (i) Town route: mean time = 20 minutes, standard deviation = 4.60 minutes
Country route: mean time = 20 minutes, standard deviation = 1.41 minutes

(ii) Both routes have the same mean time but the country route is less variable or more consistent.

8 (i) Yes. The value is more than two standard deviations above the mean rainfall.

(ii) No. The value is less than one standard deviation below the mean rainfall.

(iii) Overall mean rainfall = 1.62 cm, overall standard deviation = 0.135 cm

(iv) 84.4 cm

9 (i) No. The harvest was less than two standard deviations above the expected value.

(ii) The higher yield was probably the result of the underlying variability but that is likely to be connected to different weather patterns.

10 (i) Sample mean = 2.075 mm, standard deviation = 0.185 mm

(ii) The desired mean is less than 0.5 standard deviations from the observed mean so the machine setting seems acceptable.

11 (i) Standard deviation = 2.54 cm

(ii) The value 166 cm is less than two standard deviations above 162.82 (mean of Robert's data) and is less than two standard deviations below 170.4 (mean of Asha's data). Consequently it is impossible to say, without further information, which data set it belongs to.
The value 171 cm is more than three standard deviations above 162.82 and less than one standard deviation above 170.4 so it seems likely that this value is from Asha's data set.

12 (i) Total weight = 3147.72 g

(ii) $\sum x^2 = 84\,509.9868$

(iii) $n = 200$, $\sum x = 5164.84$, $\sum x^2 = 136\,549.913$

(iv) Mean = 25.8 g, standard deviation = 3.98 g

13 (i) 5.83 kg

(ii) 1.46 kg

14 (i) 44.1 years
(ii) 14.0 years

15 (i) Since the standard deviation is $0, all Fei's rides must cost the same. Since the mean is $2.50, it follows that both the roller coaster and the water slide cost $2.50.
(ii) $1.03

Exercise 1F (Page 47)

1 Mean = 252 g, standard deviation = 5.14 g

2 (i) $\bar{x} = -1.7$, standard deviation = 3.43
(ii) Mean = 94.83 mm, standard deviation = 0.343 mm
(iii) -18 is more than four standard deviations below the mean value.
(iv) New mean = 94.86 mm, new standard deviation = 0.255 mm

3 (i) Mean = 6.19, standard deviation = 0.484
(ii) 6.68

4 (i) No unique mode (5 & 6), mean = 5, median = 5
(ii) 50 & 60, 50, 50
(iii) 15 & 16, 15, 15
(iv) 10 & 12, 10, 10

5 (i) $\bar{x} = -4.3$ cm, standard deviation = 14.0 cm
(ii) -47 is many more than two standard deviations from the mean.
(iii) -2.05 cm, 10.2 cm

6 (i) $n = 14$

4 | 7 represents 47

```
4 | 7 9
5 | 9
6 | 2 6 7 8 8
7 | 0 2 3 4
8 | 0 4
```

Some negative skew, but otherwise a fairly symmetrical shape.
(ii) Mean = 67.1; standard deviation = 9.97

7 Mean = 6, standard deviation = 3

8 Mean = 4, standard deviation = 6.39

9 $a = 5$, standard deviation = 3.93

10 $a = 7$, standard deviation = 2.51

11 Mean = 33.75 minutes, standard deviation = 2.3 minutes

12 (i) $a = 12$
(ii) Standard deviation = 8.88

Chapter 2

❓ (Page 53)

170 friends; 143 friends (to 3 s.f.)

❓ (Page 55)

The modal class is that with the highest frequency density and so it has the tallest bar on a histogram.

❓ (Page 58)

The first interval has width 9.5, the last 10.5. All the others are 10. The reason for this is that the data can neither be negative nor exceed 70. So even if part marks were given, and so a mark such as 22.6 was possible, a student still could not obtain less than 0 or more than the maximum of 70.

Exercise 2A (Page 60)

1 (i) $0.5 \leq d < 10.5, 10.5 \leq d < 15.5,$
$15.5 \leq d < 20.5, 20.5 \leq d < 30.5,$
$30.5 \leq d < 50.5$

(ii)

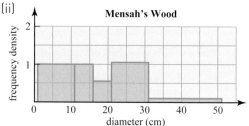

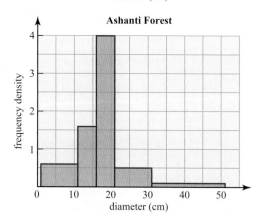

(iii) Mensah's Wood: 21–30; Ashanti Forest: 16–20

(iv) For Mensah's Wood there is a reasonably even spread of trees with diameter from 0.5 cm to 30.5 cm. For Ashanti Forest the distribution is centred about trees with diameter in the 16–20 cm interval. Neither wood has many trees with diameter greater than 30 cm.

2 (i)

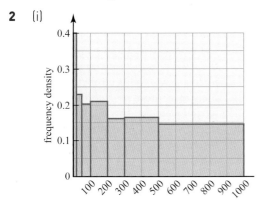

(ii) The distribution has strong positive skew.

3 (i)

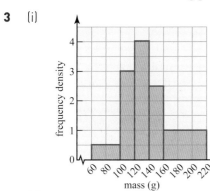

(ii) 138 g

4 (i)

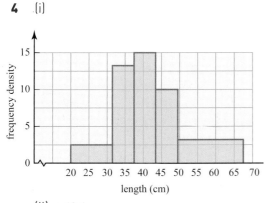

(ii) 42.1 cm

5 (i)

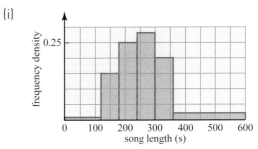

(ii) 264 seconds

6 (i) It is not clear in which interval a weight of, say, 2.5 kg should be recorded. The table does not allow for very small or very large babies. The intervals are wide so there is a risk of losing important information.

(ii) The majority of newborns are within the usual weight range. There are a few babies that are large and may cause issues during the birth process. There are many babies born with a low (below 2.5 kg) birth weight.

(iii) The doctors may need to know whether this trend is similar to other hospitals in similar areas. There may be other medical factors such as multiple births (twins and triplets), whether there are links to other diseases, whether these are first-born babies.

7 (i) The data are discrete. 'Number of pages' can only take integer values.

(ii) The data are positively skewed. Even though the data are discrete (suggesting a stem-and-leaf diagram or vertical line graph) the data are very spread out with most of the data values less than 200. A histogram will show the distribution properties best.

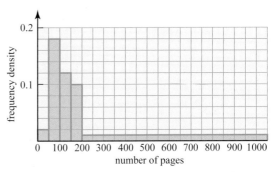

8 (i)

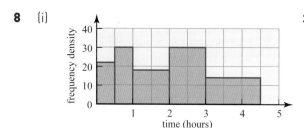

(ii) 2.1 hours

Activity 2.1 (Page 64)

For a list of n items of data, an *Excel* spreadsheet uses the 'method of hinges'. It places the median, Q_2, at position $\frac{n+1}{2}$, the lower quartile, Q_1, at position

$\frac{1}{2}\left(1 + \frac{n+1}{2}\right) = \frac{1}{2} + \frac{n+1}{4}$ and the upper quartile, Q_3, at

position $\frac{1}{2}\left(\frac{n+1}{2} + n\right) = \frac{3(n+1)}{4} - \frac{1}{2}$.

Whilst the quartiles Q_1 and Q_3 differ from those obtained with a graphical calculator, either method is acceptable.

❓ (Page 64)

The data are a sample from a parent population. The true values for the quartiles are those of the parent population, but these are unknown.

Exercise 2B (Page 72)

1 (i)   (a) 7   (b) 6
           (c) 4.5, 7.5   (d) 3
           (e) none
       (ii) (a) 15   (b) 11
           (c) 8, 14   (d) 6
           (e) none
       (iii) (a) 23   (b) 26
           (c) 23, 28   (d) 5
           (e) 14, 37
       (iv) (a) 36   (b) 118
           (c) 115, 125   (d) 10
           (e) 141

2 (i)   5, 5
       (ii) 35, 5
       (iii) 50, 50
       (iv) 80, 50

3 (i)   74
       (ii) 73, 76
       (iii) 3
       (iv)

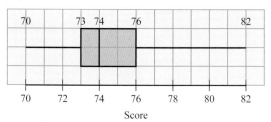

Score

(v) On average the golfers played better in the second round; their average score (median) was four shots better. However, the wider spread of data (the IQR for the second round was twice that for the first) suggests some golfers played very much better but a few played less well.

4 (i)

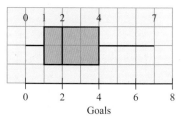

Goals

(ii) The vertical line graph as it retains more data for this small sample.

5 (i)

$y \leqslant$	Cumulative frequency
50	1
60	6
70	13
80	17
90	19
100	20

(ii)

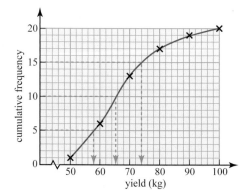

yield (kg)

(iii)   65 kg, 16 kg

(iv)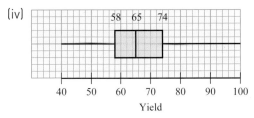
Yield

(v)   66 kg, 17.5 kg; the estimated values are quite close to these figures.

(vi)   Grouping allows one to get an overview of the distribution but in so doing you lose detail.

6 (i)

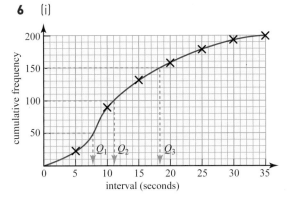
interval (seconds)

(ii)   (a)   11 s   (b)   10.5 s

(iii)   13.1 s

7

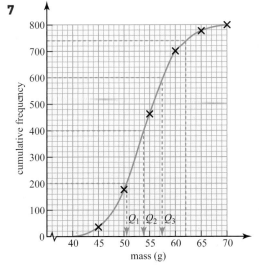
mass (g)

(i)   7.5%

(ii)   54 g

(iii)   7 g

8 (i)   12.2 s, 6.11 s

(ii)

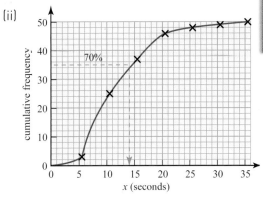
x (seconds)

14.5 seconds

9 (i) $a = 494, b = 46$

(ii)

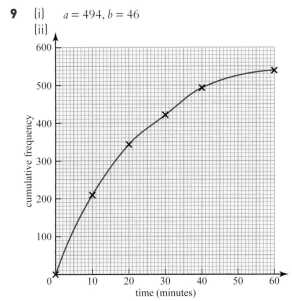
time (minutes)

(iii)   13.5 to 14.6 minutes

(iv) $m = 18.2$ minutes, $s = 14.2$ minutes

(v)   155 to 170 people

10 (i)   1 | 2 represents 12 people

    0 | 2 5 6 8 8
    1 | 2 4 6 7 7 9
    2 | 1 2 3 3 3 5 6 7
    3 | 1 5

(ii)   Median = 19, $Q_1 = 10$, $Q_3 = 24$, IQR = 14

(iii)   The median is preferable. The mode (23) is not near the centre of the data.

Chapter 3

❓ (Page 78)

See text that follows.

Exercise 3A (Page 87)

1 $\frac{66}{534}$, assuming each faulty torch has only one fault.

2 (i) $\frac{1}{6}$ (ii) $\frac{3}{6} = \frac{1}{2}$ (iii) $\frac{3}{6} = \frac{1}{2}$ (iv) $\frac{3}{6} = \frac{1}{2}$

3 (i) $\frac{12}{98} = \frac{6}{49}$ (ii) $\frac{53}{98}$ (iii) $\frac{45}{98}$ (iv) $\frac{42}{98} = \frac{3}{7}$

   (v) $\frac{56}{98} = \frac{4}{7}$ (vi) $\frac{5}{98}$

4 (i) 0.35
   (ii) The teams might draw.
   (iii) 0.45
   (iv) 0.45

5 (i)

18 *E* 12	13 3 *O* 15 ℰ
2 *S*	
10 4 1 5	
16 9 19	
6 7	
20 14 8 11 17	

   (ii) (a) $\frac{10}{20} = \frac{1}{2}$ (b) $\frac{4}{20} = \frac{1}{5}$ (c) $\frac{10}{20} = \frac{1}{2}$

      (d) $\frac{2}{20} = \frac{1}{10}$ (e) $\frac{12}{20} = \frac{3}{5}$ (f) 0

      (g) 1
      (h) $P(E \cup S) = P(E) + P(S) - P(E \cap S)$
      (i) $P(E \cup O) = P(E) + P(O) - P(E \cap O)$

Exercise 3B (page 93)

1

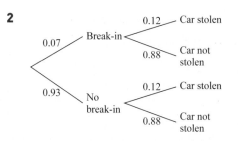

   (i) 0.2401
   (ii) 0.5002
   (iii) 0.4998

2

(tree diagram: 0.07 Break-in → 0.12 Car stolen, 0.88 Car not stolen; 0.93 No break-in → 0.12 Car stolen, 0.88 Car not stolen)

(i) 0.0084
(ii) 0.1732
(iii) 0.1816

3 (i)

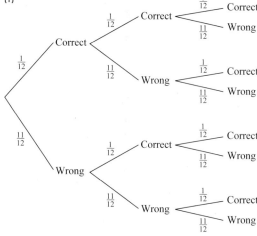

(a) 0.000579
(b) 0.770
(c) 0.0197

(ii)

(tree diagram)

(a) 0.00521
(b) 0.515
(c) 0.0938

4 0.93

5

(tree diagram: 0.008 Colour-blind → 0.2 Left-handed, 0.8 Not left-handed; 0.992 Not colour-blind → 0.2 Left-handed, 0.8 Not left-handed)

(i)   0.0016

(ii) 0.0064

(iii) 0.2064

(iv) 0.7936

6 (i) $\frac{2}{27}$ or 0.0741

(ii) $\frac{125}{216}$ or 0.579

(iii) $\frac{5}{9}$ or 0.556

7 For a sequence of events you multiply the probabilities. However, $\frac{1}{6} \times \frac{1}{6} \times \frac{1}{6} \times \frac{1}{6} \times \frac{1}{6} \times \frac{1}{6}$ gives the probability of six 6s in six throws. To find the probability of at least one 6 you need $1 - $ P(no 6s) and that is $1 - \frac{5}{6} \times \frac{5}{6} \times \frac{5}{6} \times \frac{5}{6} \times \frac{5}{6} \times \frac{5}{6} = 0.665$.

8 (i)

	First die					
	1	**2**	**3**	**4**	**5**	**6**
1	2	3	4	5	6	7
2	3	4	5	6	7	8
3	4	5	6	7	8	9
4	5	6	7	8	9	10
5	6	7	8	9	10	11
6	7	8	9	10	11	12

(Second die labels the rows 1–6)

(ii) $\frac{3}{36} = \frac{1}{12}$

(iii) 7

(iv) The different outcomes are not all equally probable.

9 0.31

❷ (Page 99)

$P(T \mid S) = \frac{109}{169} = 0.645$

$P(T \mid S') = \frac{43}{87} = 0.494$

So $P(T \mid S) \neq P(T \mid S')$

❷ (Page 99)

T represents those who had training; *T'* those with no training: *S* those who stayed in the company; $\mathscr{E}$ all employees. *S'* is inside the $\mathscr{E}$ box but not in the *S* region.

For example, in part (i) (a), the answer is $\frac{152}{256}$. 152 is in *T* (but not in *T'*), 256 is everyone.

❷ (Page 101)

The first result was used in answering part (i) and the second result in answering part (iii).

Exercise 3C (Page 101)

1 (i)   0.6   (ii)   0.556

(iii) 0.625   (iv) 0.0467

(v) 0.321   (vi) 0.075

(vii) 0.0281   (viii) 0.00218

(ix) 0.000 95   (x)   0.48

2 (i) $\frac{35}{100} = \frac{7}{20}$ (ii) $\frac{42}{100} = \frac{21}{50}$ (iii) $\frac{15}{65} = \frac{3}{13}$

3 (i) $\frac{1}{6}$ (ii) $\frac{5}{12}$ (iii) $\frac{2}{5}$

4 (i)   0.5 and 0.875

(ii) $P(B \mid A) \neq P(B)$ and $P(A \mid B) \neq P(A)$ so the events *A* and *B* are not independent.

5 (i)

	Hunter dies	Hunter lives	Total
Quark dies	$\frac{1}{12}$	$\frac{5}{12}$	$\frac{1}{2}$
Quark lives	$\frac{1}{6}$	$\frac{1}{3}$	$\frac{1}{2}$
Total	$\frac{1}{4}$	$\frac{3}{4}$	1

(ii) $\frac{1}{12}$ (iii) $\frac{5}{12}$ (iv) $\frac{5}{6}$

6 (i)

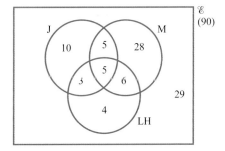

Key: J = Juniors

M = Males

LH = Left-handed players

(ii) (a) $\frac{1}{4}$ (b) $\frac{1}{6}$ (c) $\frac{28}{45}$

(d) $\frac{4}{5}$ (e) $\frac{19}{24}$ (f) $\frac{10}{39}$

7 (i) $\dfrac{618}{1281} = \dfrac{206}{427}$

(ii) $\dfrac{412}{1281}$

(iii) $P(E) = \dfrac{717}{1281} = \dfrac{239}{427}$

If M and E are independent events then
$P(M \text{ and } E) = P(M) \times P(E)$.

However, $\dfrac{412}{1281} \neq \dfrac{618}{1281} \times \dfrac{717}{1281}$ and so M
and E are not independent events.

(iv) $\dfrac{358}{564} = \dfrac{179}{282}$

8 (i) 0.4375

(ii) 0.3

9 (i) 0.252

(ii) 0.440

10 (i)

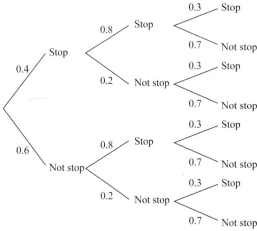

(ii) 0.224

(iii) 0.392

(iv) 0.633

11 (i) 0.15

(ii) 0.427

12 (i) 0.184

(ii) $\dfrac{1}{7}$

Chapter 4

❓ (Page 108)

You could conduct a traffic survey at peak times, over fixed periods of time, for example, 1 hour in the morning and 1 hour in the evening, over a period of a working week. You could count both the number of vehicles and the number of people travelling in each vehicle.

❓ (Page 109)

A discrete frequency distribution is best illustrated by a vertical line chart.

Using such a diagram you can see that the frequency distribution is positively skewed, see Figure 4.1.

Exercise 4A (Page 114)

1 (i)

r	$P(X = r)$
2	$\dfrac{1}{36}$
3	$\dfrac{2}{36}$
4	$\dfrac{3}{36}$
5	$\dfrac{4}{36}$
6	$\dfrac{5}{36}$
7	$\dfrac{6}{36}$
8	$\dfrac{5}{36}$
9	$\dfrac{4}{36}$
10	$\dfrac{3}{36}$
11	$\dfrac{2}{36}$
12	$\dfrac{1}{36}$

(ii)

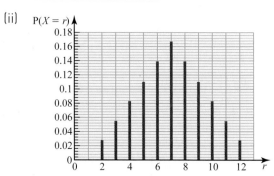

The distribution is symmetrical.

(iii) (a) $\dfrac{5}{18}$ (b) $\dfrac{1}{2}$ (c) $\dfrac{2}{3}$

2 (i)

r	0	1	2	3	4	5
$P(Y = r)$	$\dfrac{3}{18}$	$\dfrac{5}{18}$	$\dfrac{4}{18}$	$\dfrac{3}{18}$	$\dfrac{2}{18}$	$\dfrac{1}{18}$

(ii)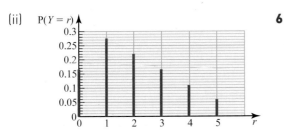

The distribution has positive skew.

(iii) (a) $\frac{2}{3}$ (b) $\frac{1}{2}$

3 (i) $k = 0.4$

r	2	4	6	8
$P(X = r)$	0.1	0.2	0.3	0.4

(ii) (a) 0.3   (b) 0.35

4 (i) $k = \frac{20}{49}$

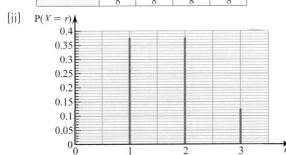

(ii) 0.248 (to 3 s.f.) $\approx$ 0.25

5 (i)

r	0	1	2	3
$P(X = r)$	$\frac{1}{8}$	$\frac{3}{8}$	$\frac{3}{8}$	$\frac{1}{8}$

(ii)

P(X = r)

The distribution is symmetrical.

(iii) $\frac{1}{2}$

(iv) No, with four coins, an equal number of heads and tails is possible, so probability $< \frac{1}{2}$.

6

r	$P(X = r)$
1	$\frac{1}{16}$
2	$\frac{2}{16}$
3	$\frac{2}{16}$
4	$\frac{3}{16}$
6	$\frac{2}{16}$
8	$\frac{2}{16}$
9	$\frac{1}{16}$
12	$\frac{2}{16}$
16	$\frac{1}{16}$

(ii) $\frac{1}{4}$

7 (i) $k = 0.08$

r	0	1	2	3	4
$P(X = r)$	0.2	0.24	0.32	0.24	0

(ii) Let Y represent the number of chicks.

r	0	1	2	3
$P(Y = r)$	0.351 04	0.449 28	0.184 32	0.015 36

8 (i) $a = 0.42$

(ii) $k = \frac{1}{35}$

(iii)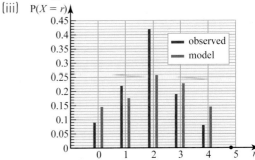

Since the probability distributions look quite different, the model is not a good one.

9 (i) $P(X = 1) = \frac{1}{216}$

(ii) $P(X \leqslant 2) = \frac{8}{216} = \frac{1}{27}$

(iii) $P(X \le 3) = \frac{27}{216}$, $P(X = 3) = \frac{19}{216}$;

$P(X \le 4) = \frac{64}{216}$, $P(X = 4) = \frac{37}{216}$;

$P(X \le 5) = \frac{125}{216}$, $P(X = 5) = \frac{61}{216}$;

$P(X \le 6) = 1$, $P(X = 6) = \frac{91}{216}$

(iv)

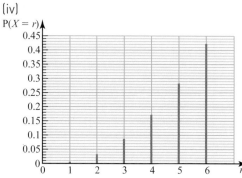

The distribution has negative skew.

10 (i)

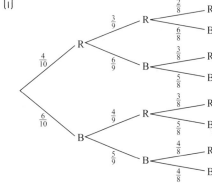

(ii)

r	0	1	2	3
P(X = r)	$\frac{1}{6}$	$\frac{1}{2}$	$\frac{3}{10}$	$\frac{1}{30}$

11 (i) $\frac{3}{11}$

(iii)

r	0	1	2	3
P(X = r)	$\frac{14}{55}$	$\frac{28}{55}$	$\frac{12}{55}$	$\frac{1}{55}$

❓ (Page 116)

Car Share World could have based their claim on the results of traffic surveys. These could be used to calculate summary statistics such as the mean and standard deviation of the number of people per vehicle, as well as the number of vehicles per hour.

❓ (Page 117)

By comparing measures of central tendency and spread, it is possible to infer whether or not there is a significant difference between their values. There are many different tests of statistical inference; you will learn about these if you go on to study *Probability & Statistics 2*. It is also possible to compare statistically the proportion of vehicles with a single occupant.

Activity 4.1 (Page 117)

Mean = 2
Variance = 0.86

Activity 4.2 (Page 119)

Mean = 1.7
Variance = 0.87
The increase in average occupancy, together with a significant reduction in the proportion of vehicles with a single occupant, could be used to infer that the scheme has been successful.
The two measures of spread are almost the same.

❓ (Page 120)

If the expectation, E(X), is not exact in decimal form, then calculations by hand using the definition of Var(X) may be tedious and/or prone to arithmetic errors by premature approximation of E(X). The alternative formulation of Var(X) may be more appropriate in such cases.

Exercise 4B (Page 121)

1 $E(X) = 3$

2 (i) $E(X) = 3.125$
(ii) $P(X < 3.125) = 0.5625$

3 (i) $p = 0.8$

r	4	5
P(X = r)	0.8	0.2

(ii)

r	50	100
P(Y = r)	0.4	0.6

4 (i) $E(X) = 7$
(ii) $Var(X) = 5.83$
(iii) (a) $\frac{5}{12}$ (b) $\frac{1}{6}$ (c) $\frac{17}{18}$

5 (i) $E(X) = 1.94; \text{Var}(X) = 2.05$

   (ii) (a) $\dfrac{5}{9}$ (b) $\dfrac{1}{18}$

6 (i) $E(X) = 1.5$
   (ii) $\text{Var}(X) = 0.75$
   (iii) $E(Y) = 5, \text{Var}(Y) = 2.5$

7 (i) $k = 0.1$
   (ii) 1.25
   (iii) 0.942 (to 3 s.f.)

8 (ii)

r	0	1	2	3	4	6
$P(X = r)$	$\dfrac{1}{4}$	$\dfrac{1}{3}$	$\dfrac{1}{9}$	$\dfrac{1}{6}$	$\dfrac{1}{9}$	$\dfrac{1}{36}$

   (iii) $E(X) = \dfrac{5}{3}, \text{Var}(X) = \dfrac{41}{18}$

9 (i)

r	0	80	120	160
$P(X = r)$	$\dfrac{5}{28}$	$\dfrac{20}{28}$	$\dfrac{2}{28}$	$\dfrac{1}{28}$

   (ii) $E(X) = \$71.43, \text{Var}(X) = 1410$ (to 3 s.f.)
   (iii) $E(W) = \$500, \text{Var}(X) = 3600 \text{ or } 13\,200$

10 $a = 0.2, b = 0.25$

11 (i)

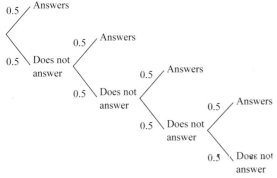

Attempt 1   Attempt 2   Attempt 3   Attempt 4

   (ii)

x	0	1	2	3	4
$P(X = r)$	$\dfrac{1}{2}$	$\dfrac{1}{4}$	$\dfrac{1}{8}$	$\dfrac{1}{16}$	$\dfrac{1}{16}$

   (iii) $\dfrac{15}{16}$

12 (ii) $E(X) = 3.75, \text{Var}(X) = 2.1875$

13 (i) $2q - p = 0.39, p + q = 0.42; p = 0.15,$
              $q = 0.27$

   (ii) 2.5875

14 (i) $\dfrac{1}{6}$

   (ii) $E(X) = \dfrac{4}{3}, \text{Var}(X) = 7\dfrac{5}{9}$

15 (i) 0.195

   (ii)

r	2	3	4	5	6	7
$P(X = r)$	$\dfrac{1}{36}$	0	$\dfrac{1}{18}$	0	$\dfrac{5}{36}$	$\dfrac{1}{9}$

r	8	9	10	11	12
$P(X = r)$	$\dfrac{1}{9}$	$\dfrac{1}{9}$	$\dfrac{1}{9}$	$\dfrac{2}{9}$	$\dfrac{1}{9}$

   (iii) $8\dfrac{2}{3}$

   (iv) $\dfrac{5}{9}$

Chapter 5

❓ (Page 126)

See text that follows.

❓ (Page 127)

Oscar could have put the bricks in order by chance. A probability of $\dfrac{1}{120}$ is small but not very small. What would really be convincing is if he could repeat the task whenever he was given the bricks.

Exercise 5A (Page 131)

1 (i) $40\,320$
   (ii) 56
   (iii) $\dfrac{5}{7}$

2 (i) $\dfrac{1}{n}$
   (ii) $n - 1$

3 (i) $(n + 3)(n + 2)$
   (ii) $n(n - 1)$

4 (i) $\dfrac{8! \times 2!}{5! \times 5!}$

   (ii) $\dfrac{16!}{14! \times 4!}$

   (iii) $\dfrac{(n + 1)!}{(n - 2)! \times 4!}$

5 (i) $9 \times 7!$
   (ii) $n!(n + 2)$

6 24
7 $40\,320$
8 720

9 (i) 120
   (ii) $\dfrac{1}{120}$

10 (i) 14!

    (ii) $\dfrac{1}{14!}$

11 (i) $\dfrac{1}{10!}$

    (ii) 1.38 (i.e. 1 or 2 people)

12 $\dfrac{6}{24} = \dfrac{1}{4}$ (OPTS, POST, POTS, SPOT, STOP, TOPS)

13 (i) 24
    (ii) 120
    (iii) 362 880
    (iv) 12
    (v) 420
    (vi) 50 400

14 (i) 2520
    (ii) 360
    (iii) 720
    (iv) 1800

Investigations (Page 132)

1 $n > 6, n > 7, n > 8$

2 Examples are $\dfrac{7!}{5!}$, $3! + 3! + 3! + 3! + 3! + 3! + 3!$, $4! + 4! - 3!$, etc.

3 (i) 4^4
    (ii) (a) 12
        (b) 4

❓ (Page 133)

No, it does not matter.

❓ (Page 134)

Multiply top and bottom by 43!

$$\dfrac{49 \times 48 \times 47 \times 46 \times 45 \times 44}{6!} \times \dfrac{43!}{43!}$$

$$= \dfrac{49!}{6 \times 43!}$$

❓ (Page 134)

$${}^{49}C_6 = \begin{pmatrix} 49 \\ 6 \end{pmatrix} = 13\,983\,816 \approx 14 \text{ million}$$

❓ (Page 134)

By following the same argument as for the example of the maths teacher's game, but with n for 49 and r for 6.

❓ (Page 135)

$${}^{n}C_0 = \begin{pmatrix} n \\ 0 \end{pmatrix} = \dfrac{n!}{0! \times n!} = 1 \text{ if } 0! = 1$$

$${}^{n}C_n = \begin{pmatrix} n \\ n \end{pmatrix} = \dfrac{n!}{n!\,(n - n)!} = 1 \qquad \text{again if } 0! = 1$$

❓ (Page 138)

The probability is $\dfrac{1}{24}$, assuming the selection is done at random, so R Chowdhry is not justified in saying 'less than one in a hundred'.

As a product of probabilities $\dfrac{1}{3} \times \dfrac{1}{2} \times \dfrac{1}{4} = \dfrac{1}{24}$

Exercise 5B (Page 140)

1 (i) (a) 30   (b) 1680
     (c) 5040
    (ii) (a) 15   (b) 70
     (c) 210

2 $\dfrac{1}{2730}$

3 715

4 280

5 (i) 31 824
    (ii) 3000

6 (i) 210
    (ii) (a) $\dfrac{1}{14}$ (b) $\dfrac{3}{7}$

7 (i) 126
    (ii) (a) $\dfrac{1}{126}$ (b) $\dfrac{45}{126}$

8 (i) $\dfrac{1}{120}$

    (ii) $\dfrac{1}{7\,893\,600}$

9 (i) (a) 15   (b) 75
    (ii) (a) 90 720 (b) 120

10 (a) (i) 15 120
      (ii) 10 080
    (b) 47 880

11 (i) 4.94×10^{11}
    (ii) 79 833 600
    (iii) 21

12 (i) 2 177 280
    (ii) 90

13 (i) 33 033 000
    (ii) 86 400
    (iii) 288

14 (i) 259 459 200
    (ii) 3 628 800
    (iii) 0.986

15 (i) 831 600
    (ii) 900
    (iii) 126

16 (a) (i) 60 (ii) 216
    (b) (i) 1316 (ii) 517

Chapter 6

❷ (Page 146)

See text that follows.

❷ (Page 146)

She should really try to improve her production process so as to reduce the probability of a bulb being substandard.

Exercise 6A (Page 149)

1 $\dfrac{15}{64}$

2 0.271

3 0.214

4 0.294

5 (i) 0.146
    (ii) Poor visibility might depend on the time of day, or might vary with the time of year. If so, this simple binomial model would not be applicable

6 (i) $\dfrac{1}{8}$
    (ii) $\dfrac{3}{8}$
    (iii) $\dfrac{3}{8}$
    (iv) $\dfrac{1}{8}$

7 (i) 0.246
    (ii) Exactly 7 heads

8 (i) (a) 0.0576 (b) 0.198
      (c) 0.296 (d) 0.448
    (ii) 2

9 (i) (a) 0.264 (b) 0.368
      (c) 0.239 (d) 0.129
    (ii) Assumed the probability of being born in January $= \dfrac{31}{365}$. This ignores leap years and the possibility of seasonal variations in the pattern of births throughout the year.

10 The three possible outcomes are not equally likely: 'one head and one tail' can arise in two ways (HT or TH) and is therefore twice as probable as 'two heads' (or 'two tails').

Activity 6.1 (Page 152)

Expectation of $X = \displaystyle\sum_{r=0}^{n} r \times P(X = r)$

Since the term with $r = 0$ is zero

Expectation of $X = \displaystyle\sum_{r=1}^{n} r \times P(X = r)$

$$= \sum_{r=1}^{n} r \times {}^{n}C_{r}\, p^{r} q^{n-r}.$$

The typical term of this sum is

$$r \times \frac{n!}{r!(n-r)!}\, p^{r} q^{n-r} = np \times \frac{(n-1)!}{(r-1)!(n-r)!}\, p^{r-1} q^{n-r}$$

$$= np \times \frac{(n-1)!}{(r-1)!((n-1)-(r-1))!}$$

$$\times p^{r-1} q^{(n-1)-(r-1)}$$

using $(n-1) - (r-1) = n - r$

$$= np \times {}^{n-1}C_{r-1}\, p^{r-1} q^{(n-1)-(r-1)}$$

$$= np \times {}^{n-1}C_{s}\, p^{s} q^{(n-1)-s}$$

where $s = r - 1$

In the summation, np is a common factor and s runs from 0 to $n - 1$ as r runs from 1 to n. Therefore

Expectation of $X = np \times \displaystyle\sum_{s=0}^{n-1} {}^{n-1}C_{s}\, p^{s} q^{(n-1)-s}$

$$= np(q + p)^{n-1}$$

$$= np \qquad \text{since } q + p = 1.$$

Exercise 6B (Page 154)

1 (i) (a) 0.000 129 (b) 0.0322 (c) 0.402
    (ii) 0 and 1 are equally likely

2 (i) 2
    (ii) 0.388
    (iii) 0.323

3 (i) (a) 2.5 (b) 5 (c) 7.5
    (ii) (a) 1.875 (b) 2.5 (c) 1.875

4 (i)   0.240
     (ii) 0.412
     (iii) 0.265
     (iv) 0.512
     (v)   0.384
     (vi) 0.096
     (vii) 0.317
           Assumption: the men and women in the office are randomly chosen from the population (as far as their weights are concerned).

5 (i)   (a) $\frac{1}{81}$ (b) $\frac{8}{81}$ (c) $\frac{24}{81}$ (d) $\frac{32}{81}$
     (ii) 2 min 40 s

6 (i)   0.0735
     (ii) Mean = 2.7, variance = 1.89
     (iii) 3
     (iv) 0.267

7 (i)   constant/given p, independent trials, fixed/given number of trials, only two outcomes
     (ii) 0.520

8 0.655

9 (i)   0.994
     (ii) 0.405

10 (i)   0.132
     (ii) 0.0729
     (iii) 0.0100
     (iv) Mean = $\frac{5}{3}$, variance = $\frac{10}{9}$

11 0.212

Exercise 6C (Page 160)

1 (i) $\frac{1}{6}$
     (ii) $\frac{25}{216}$
     (iii) $\frac{11}{36}$
     (iv) $\frac{125}{216}$

2 (i) $\frac{1}{3}$
     (ii) $\frac{16}{243}$
     (iii) $\frac{32}{243}$

3 (i)   (a) $\frac{256}{625}$
           (b) $\frac{369}{625}$

     (c) $\frac{1024}{3125}$
     (d) $\frac{256}{3125}$
     (ii) Answers (b), (c) and (d) add to 1. They cover all possible outcomes.

4 (i) $\frac{1}{6}$
     (ii) 6
     (iii) $\frac{3125}{46656}$

5 (i) $\frac{243}{1024}$
     (ii) $\frac{781}{1024}$
     (iii) $\frac{81}{1024}$
     (iv) $\frac{27}{1024}$

6 (i)   0.2
     (ii) 0.0819
     (iii) 0.410
     (iv) 5
     (v)   4

7 (i)   0.0791
     (ii) 0.237
     (iii) 0.4375
     (iv) 4

8 (i)   0.0504
     (ii) 0.168
     (iii) $3\frac{1}{3}$
     (iv) 0.123

9 (i) $\frac{32}{243}$
     (ii) $\frac{16}{243}$
     (iii) $\frac{5}{3}$
     (iv) $3\frac{1}{3}$

10 (i)   Geo$\left(\frac{1}{6}\right)$
     (ii) (a) 0.0965
           (b) 0.421
           (c) 0.482
     (iii) $\frac{3}{25}$
     (iv) (a) 0.482
           (b) 0.0934

Chapter 7

❷ (Page 164)

See text that follows.

Exercise 7A (Page 175)

1 (i)   0.841
   (ii) 0.159
   (iii) 0.5
   (iv) 0.341

2 (i)   0.726
   (ii) 0.274
   (iii) 0.0548
   (iv) 0.219

3 (i)   0.159
   (ii) 0.841
   (iii) 0.841
   (iv) 0.683

4 (i)   0.841
   (ii) 0.0228
   (iii) 0.136

5 (i)   0.0668
   (ii) 0.691
   (iii) 0.242

6 (i)   0.0668
   (ii) 0.159
   (iii) 0.775

7 (i)   31 g, 1.97 g
   (ii) 2 or 3, 13, 34 or 35, 34 or 35, 13, 2 or 3 eggs
   (iii) More data would need to be collected to say reliably that the weights are normally distributed.

8 (i)   78.7%
   (ii) 5.25 cm, 0.0545 cm

9 $\mu = 54.1$ and $\sigma = 2.88$

10 0.537

11 $\mu = 2.34$ and $\sigma = 7.91$

12 (i)   30 or 31 sheep
    (ii) 73.6
    (iii) 64.2

13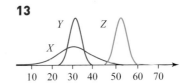

14 (i)   9.14
    (ii) 0.159

15 (i)   0.00430
    (ii) 1.71 bars, 2.09 bars

16 (i)   7.30 cm
    (ii) 0.136
    (iii) 0.370

17 (a)   7.24
    (b)   546

18 (i)   8.75
    (ii) 0.546

19 (i)   0.595
    (ii) 0.573

20 (i)   0.0276
    (ii) 7.72°C

21 (i)   0.484
    (ii) 96.9 minutes,
          103.1 minutes

❷ (Page 183)

One possibility is that some people, knowing their votes should be secret, resented being asked who they had supported and so deliberately gave wrong answers. Another is that the exit poll was taken at a time of day when those voting were unrepresentative of the electorate as a whole.

Exercise 7B (Page 183)

1 (i)   0.282
   (ii) 0.526
   (iii) 25, 4.33
   (iv) 0.102

2 0.704

3 (i) $\frac{2}{9}$
   (ii) $\frac{4}{243}$
   (iii) 0.812

4 0.677

5 (i) 0.126
     (ii) 0.281

6 (i) 0.748
     (ii) 0.887

7 (i) 0.298
     (ii) 0.118
     (iii) 13

8 (i) 0.311
     (ii) Not appropriate because $np < 5$.
     (iii) 0.181

Index